SpringerBriefs in Molecular Science

Electrical and Magnetic Properties of Atoms,
Molecules, and Clusters

Series editor

George Maroulis, Patras, Greece

More information about this series at http://www.springer.com/series/11647

Feng Long Gu · Yuriko Aoki
Michael Springborg · Bernard Kirtman

Calculations on nonlinear optical properties for large systems

The elongation method

 Springer

Feng Long Gu
School of Chemistry and Environment
South China Normal University
Guangzhou, Guangdong
China

Yuriko Aoki
Department of Energy and Material
  Sciences
Kyushu University
Kasuga, Fukuoka
Japan

Michael Springborg
Physical and Theoretical Chemistry
University of Saarland
Saarbrücken
Germany

Bernard Kirtman
Department of Chemistry and Biochemistry
University of California
Santa Barbara, CA
USA

ISSN 2191-5407   ISSN 2191-5415   (electronic)
ISBN 978-3-319-11067-7   ISBN 978-3-319-11068-4   (eBook)
DOI 10.1007/978-3-319-11068-4

Library of Congress Control Number: 2014956657

Springer Cham Heidelberg New York Dordrecht London

Printed on acid-free paper

Springer is part of Springer Science+Business Media (www.springer.com)

*Anyone who has never made a mistake
has never tried anything new*

—Albert Einstein

Preface

There has been a long-term interest in nonlinear optical (NLO) phenomena, especially as they may be utilized to elucidate physicochemical properties and, more recently, for the development of practical devices in many different fields. A wide variety of processes occurring in either the weak field (perturbative) or strong field (non-perturbative) regime fall under this rubric. Systems that have been investigated range from simple small molecules to large complex bio-and nano-materials and crystalline systems. In either case, advances have been fueled by a combination of experiment and theory. Naturally, the theoretical/computational tools available for larger systems are not as far advanced, but intensive efforts are underway to develop suitable approaches.

One initial approach has been to consider systems that can be modeled as being infinite and periodic. A major computational simplification is then afforded by translational symmetry, although special methods to employ that symmetry are required. On the other hand, structural and chemical deviations from perfect stereoregularity can have important and interesting consequences. It is particularly for such cases that finite-cluster treatments, such as the elongation method (ELG), enter the picture. In general, the finite cluster and infinite periodic approaches have complementary advantages and disadvantages. While exploring the full potentiality of each separately, the possibilities for a symbiotic combination should also be borne in mind.

The elongation method is a very flexible finite cluster approach that has a number of advantageous aspects. It was originally designed to mimic the buildup of an infinite periodic polymer by addition of successive monomer units to a growing chain. However, it is not necessary for the added units to be identical nor is the buildup limited to one-dimensional or quasi-one-dimensional systems. The shape of the system, in fact, is arbitrary. In adding units the elongation method utilizes locality to create a procedure that scales linearly with the size of the system even in cases where partial delocalization must be taken into account. The efficiency depends on the method of localization and, although still evolving, highly satisfactory procedures have already been realized. Moreover, hybrid methods that include the combination with a partially periodic calculation can readily be foreseen.

This monograph is composed of six chapters. In Chap. 1 we survey the broad range of nonlinear optical properties as well as their various realized and potential applications. This leads to a discussion of nonlinear optical materials. For design purposes one needs to relate the structure of proposed materials to their nonlinear optical (and other) properties, which is a situation where theoretical approaches can be very helpful in providing suggestions for candidate systems that subsequently can be synthesized and studied experimentally. Thus, in Chaps. 2 and 3, we turn to the quantum-mechanical treatment of the response to one or more external oscillating electric fields for molecular (Chap. 2) and macroscopic crystalline (Chap. 3) systems. The systems that can be efficiently studied with the elongation method lie somewhere between these two classes of materials. Thus, the molecular and crystalline systems provide limiting cases whose theoretical treatment can give additional, useful information.

In Chaps. 2 and 3 we present two different ways of including the electric field(s), i.e., by means of the scalar potential or of the vector potential. Whereas our initial focus is on the effective single particle [i.e., Hartree-Fock (HF) or Kohn–Sham density functional theory (KS-DFT)] treatment of the electronic response we include the local MP2 treatment of electron correlation as well. Besides the pure electronic response there is also a nuclear response that gives rise to vibrational NLO properties. The latter can be very important, or even dominant, and cannot be ignored in evaluating the NLO properties. This vibrational contribution can be determined either by perturbation theory or, especially in large systems, by means of the so-called finite (i.e., static) field-nuclear relaxation procedure. In addition, we demonstrate how time/frequency-dependent responses can be calculated using a coupled perturbed theory both at the HF and at the KS-DFT level.

The ELG method is a set of procedures for building up a large system by adding small fragments to an original and growing cluster, one fragment at a time. In Chaps. 4 and 5 we show how this is accomplished by considering, first, the field-free problem at the single particle level. Of particular importance are the procedures for regional molecular orbital localization, as well as integral (and other) cut-offs, that act in concert to allow calculations to be restricted to an interactive region of essentially fixed size which is considerably smaller than the complete system. In carrying out calculations there will sometimes be a limited number of delocalized molecular orbitals that cannot meet the localization criteria. A provision for dealing with this circumstance is described. Building upon the single particle case, we also present our formulation of field-free ELG-LMP2 and ELG-CIS for excited states. Finally, in Chap. 5 we present the adaptations that are utilized in the finite field and, particularly, the coupled perturbation theory treatments (HF; KS-DFT) of NLO properties.

By its very nature the ELG methodology leads to the prediction of linear scaling, which is shown to be verified in practice. Chapters 4 and 5 contain a variety of calculations that establish the efficiency, accuracy, and scope of the ELG approach. The first set of calculations concerns quasilinear systems including conjugated polymers and chains of water or benzene molecules. Also, ring systems containing porphyrins are discussed. In addition to demonstrating that the computational costs scale linearly with system size, the accuracy of field-free, finite field, and NLO properties is established by comparing with results of conventional molecular calculations. These studies include tests of the ELG treatment of delocalized molecular orbitals. The second set of calculations is presented to demonstrate the applicability of ELG to large systems that are not quasilinear. For such systems there may be significant interactions between fragments that are not directly connected in the buildup process. We show how these interactions can be taken into account within the framework of the ELG method. Moreover, we confirm the efficiency and accuracy achieved for quasilinear systems.

Although the ELG method has advanced considerably since the time of its initial proposal in the early 1990s, there is still much room for improvements and extensions pertinent to NLO as well as other properties. Some specific examples and more general suggestions are given in Chap. 6 together with a concluding summary.

The four authors meeting in Bernie's office in University of California, Santa Barbara, March 15, 2013. From *left to right* Feng Long Gu, Michael Springborg, Yuriko Aoki, Bernard Kirtman

June 2014

Feng Long Gu
Yuriko Aoki
Michael Springborg
Bernard Kirtman

Acknowledgments

We would express our great acknowledgments to Prof. Akira Imamura for his encouragement and his original idea of the elongation method, an efficient method for quantum mechanical calculations of large systems, awarded by Japan Science and Technology Agency (JST)-PRESTO Sakigake21 (1991–1994), and further continued in YA's group financed by ACT-JST, JST-PRESTO and JST-CREST. The authors are also grateful to the former postdoctoral fellows and researchers under the project of the development of the elongation method: Jacek Korchowiec, Marcin Makowski, Yanliang Ren, Likai Yan, Yuuchi Orimoto, Weiquan Tian, Vladimir Pomogaev, Wei Chen, Guangtao Yu, Oleksandr Loboda, Inerbae Talgat, David Price, Liang Peng, and graduate students: Shinichi Ohnish, Anna Pomogaeva, Kai Liu, Peng Xie, Xun Zhu, Shinichi Abe, Ryota Tsutsui, Huaqing Yang, and other students, for their elaborating research on this project. Thanks are also due to Yi Dong and Jorge Vargas for their contributions to the treatment of infinite, periodic systems. Release in GAMESS program package of the elongation method was admitted with great support by Dr. Michael Schmidt at Iowa State University. The calculations were mainly performed on the linux clusters of the laboratories provided by JST-CREST in Gu's group at South China Normal University and Aoki's group at Kyushu University, as well as the high performance computing system in Research Institute for Information Technology at Kyushu University. Besides JST-CREST, this work was partly supported by a grant-in-aid from the Ministry of Education, Culture, Sports, Science and Technology (MEXT) of Japan (No. 04205104, 04453016, 07554087, 08454183, 08740548, 09740525, 14340185, 16655009, 19350012, 21655007, 23245005) and the Japan Society for the Promotion of Science (JSPS). This work is also partly supported by the National Natural Science Foundation of China (No. 21273081, No. 21073067) and Guangdong Province Universities and Colleges Pearl River Scholar Funded Scheme (2011) as well as the German Research Council (DFT; project no. Sp439/37). We thank Dr. Kai Liu and Ms. Ikuko Okawa for preparing some figures and

Ms. Linping Hu for her typing of the main part of the equations and final checking of all references. Finally, one of the authors (MS) is very grateful to the International Center for Materials Research, University of California, Santa Barbara, for generous hospitality.

Contents

Acronyms

AC	Alternating Current
AO	Atomic Orbital
BBO	β-Barium Borate
BvK	Born von Kármán
CARS	Coherent Anti-Stokes Spectroscopy
CI	Configuration Interaction
CMO	Canonical Molecular Orbital
CN	Clamped Nucleus
CNV	Conventional (in contrast to Elongation Method) Calculation
CPHF	Coupled Perturbed Hartree-Fock
CPKS	Coupled Perturned Kohn-Sham
CR	Charge Recombination
CS	Charge Separation
DC	Direct Current
DC-P	DC-Pockels
DC-SHG	DC-Second Harmonic Generation
DFT	Density Functional Theory
DFWM	Degenerate Four-Wave Mixing
EFIOR	Electric Field-Induced Optical Rectification
EFISHG	Electric Field-Induced Second Harmonic Generation
ELG	Elongation Method
EOKE	Electro-Optic Kerr Effect
EOPE	Electro-Optic Pockels Effect
EOR	Electro-Optic Rectification
F6A	Poly[(9,9-dihexylfluorene-2,7-diyl)-co-(anthracene-9,10-diyl)]
F6PC	Poly(9,9-n-dihexyl-2,7-fluorene-alt-9-phenyl-3,6-carbazole)
F8BT	Poly[(9,9-di-n-octylfluorenyl-2,7-diyl)-alt-(benzo[2,1,3]thiadiazol-4,8-diyl)]
FF	Finite Field
FF-NR	Finite-Field Nuclear-Relaxation

FIC	Field-Induced Coordinates
G-ELG	Generalized Elongation Method
IDRI	Intensity-Dependent Refractive Index
KDP	Potassium Dihydrogen Phosphate
KTP	Potassium Titanyl Phosphate
LCIS	Local Configuration Interaction Singles
LDOS	Local Density of States
LMO	Localized Molecular Orbital
LMP2	Local Møller-Plesset 2nd order Perturbation Theory
L&NLO	Linear and Nonlinear Optical
MO	Molecular Orbital
MP	Møller-Plesset Perturbation Theory
NBO	Natural Bond Orbital
NLO	Nonlinear Optical
NR	Nuclear Relaxation
OAO	Orthonormalized Atomic Orbital
OKE	Electro-Optical Kerr Effect
OR	Optical Rectification
PBC	Periodic Boundary Conditions
PCM	Polarizable Continuum Model
PSC	Polymeric Solar Cells
RLMO	Regional Localized Molecular Orbital
RO	Regional Orbitals
RPA	Random-Phase Approximation
SCF	Self-consistent Field
SFG	Sum-Frequency Generation
SHG	Second Harmonic Generation
SOS	Sum Over States
SOPPA	Second-order Polarization Propagator Approximation
SRIP	Symmetric Resonance Structures with Invertible Polarization
TDDFT	Time-Dependent Density-Functional Theory
TDHF	Time-Dependent Hartree-Fock
TDKS	Time-Dependent Kohn-Sham
THG	Third Harmonic Generation
UC	Uncoupled
VPA	Vector-Potential Approach
ZPVA	Zero-Point Vibrational Average

Chapter 1
Survey of Nonlinear Optical Materials

Abstract This chapter contains a brief summary of various NLO processes as well as materials that show particularly large responses.

1.1 Introduction

Nonlinear optics (NLO) refers to phenomena where the nonlinear dielectric polarization of a material in response to incident light causes the properties of the emergent light, such as its phase, frequency, amplitude ... to be altered. Because of the wide variety of NLO processes and their practical technological applications this field has drawn much attention from the chemistry, physics and engineering communities.

NLO phenomena are nowadays, typically, observed at very high light intensities like those provided by pulsed lasers. Thus, although some processes had already been observed much earlier, the modern era may be said to have been initiated with the report of the first (ruby) laser device by Maiman in 1960 [1]. Shortly thereafter Franken et al. used a ruby laser to demonstrate second-harmonic generation (SHG) [2]. Figure 1.1 shows a schematic illustration of their experiment wherein light of wavelength 694.30 nm from the laser is passed through a quartz sample and partially converted to light with a wavelength of 347.15 nm, which corresponds to frequency doubling, i.e. SHG.

Bloembergen et al. [3] followed the experiment by Franken et al. with a general quantum-mechanical treatment of the nonlinear susceptibilities (polarization) up to terms cubic in the electric fields and, among other things, discussed the application of their theory to SHG (quadratic in the field). Then, in a second paper Bloembergen and Pershan [4] calculated the SHG and sum-frequency generation (SFG) at surfaces. These papers laid the foundation for the perturbation theory treatment of nonlinear optical processes. Since that time a very large number of theoretical/computational studies for various NLO processes have been presented and many different types of NLO material have been investigated.

Nonlinear effects fall into two qualitatively different categories, parametric and non-parametric, each deserving a treatise of its own. In this monograph we focus on parametric nonlinear optical phenomena, which correspond to processes where the

© The Author(s) 2015   1
F.L. Gu et al., *Calculations on nonlinear optical properties for large systems*,
SpringerBriefs in Electrical and Magnetic Properties of Atoms,
Molecules, and Clusters, DOI 10.1007/978-3-319-11068-4_1

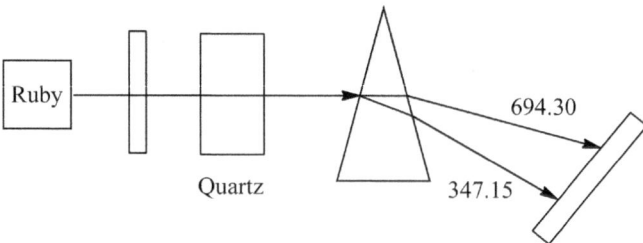

Fig. 1.1 The second-harmonic generation experiment by Franken et al. at the University of Michigan in 1961. The splitting of the incoming electric wave with a wavelength of 694.30 nm into two with different wavelengths (694.30 and 347.15 nm, respectively) is shown

dielectric polarization P_i (=ith component of the dipole moment per unit volume) induced in a macroscopic sample by electromagnetic waves (light or other electro-magnetic radiation) can be described through a Taylor series expansion in the electric field (or fields):

$$P_i(\omega_\sigma)/\varepsilon_0 = \chi_{ij}^{(1)} E_j(\pm\omega_1) + \chi_{ijk}^{(2)} E_j(\pm\omega_1) E_k(\pm\omega_2)$$
$$+ \chi_{ijkl}^{(3)} E_j(\pm\omega_1) E_k(\pm\omega_2) E_l(\pm\omega_3) + \cdots \quad (1.1)$$

In the above equation $E_j(\pm\omega_1)$, for example, is the jth Cartesian component of the electric field (assumed to be spatially uniform over the entire sample) character-ized by the time-dependent frequency factor $\exp(\pm i\omega_1 t)$. In addition to the optical field(s) a dc field ($\omega = 0$) may also be present. The particular NLO process that is observed depends primarily upon whether it is second- or third-order in the field and on the frequency, ω_σ, at which the induced dipole moment oscillates. The latter is simply the sum of the frequencies $\pm\omega_1 \pm\omega_2$ (second-order) or $\pm\omega_1 \pm\omega_2 \pm\omega_3$ (third-order). Finally, the strength of the process is governed by the corresponding expansion coefficient, $\chi_{ijk}^{(2)}$ or $\chi_{ijkl}^{(3)}$, which is the macroscopic susceptibility for that process. Note that, for convenience, the frequencies that characterize the particu-lar susceptibility have been suppressed. As an example, the case of SHG discussed above, corresponds to the monochromatic process ($\omega_1 = \omega_2 = \omega$) so that

$$P_i(2\omega) = \varepsilon_0 \chi_{ijk}^{(2)}(-2\omega; \omega, \omega) E_j(\omega) E_k(\omega) \quad (1.2)$$

where $\chi_{ijk}^{(2)}(-2\omega; \omega, \omega)$ is the ijk component of the second-order (or first non-linear) macroscopic susceptibility tensor for second harmonic generation. There is also a second-order susceptibility tensor for the dc-Pockels effect given by $\chi_{ijk}^{(2)}(-\omega; \omega, 0)$. In fact, there are other second-order processes that are of interest, both monochromatic and non-monochromatic and an even larger number of third-order processes. Some of the most important of these NLO processes will be described in detail in the next section.

Associated with each macroscopic susceptibility tensor there is a microscopic hyperpolarizability tensor, which gives the induced dipole moment for an individual molecule (or specified chemical unit) in the sample. The definition of the microscopic quantities is essentially the same as in Eq. (1.1) except that P_i/ε_0 is replaced by the dipole moment μ_i, the macroscopic field is replaced by the microscopic field (i.e. the field acting on the individual molecule) and, of course, the macroscopic susceptibilities are replaced by the (hyper)polarizabilities:

$$\chi^{(1)} \to \alpha; \quad \chi^{(2)} \to (\frac{1}{2})\beta; \quad \chi^{(3)} \to (\frac{1}{6})\gamma. \tag{1.3}$$

Here, the parentheses indicate that different notations may be found with or without the prefactors $\frac{1}{2}$ and $\frac{1}{6}$.

For design of new materials it is the hyperpolarizabilities that are normally computed to predict how the measured macroscopic NLO properties will vary from one case to another and that is the approach that will be taken here. As a base line there is also interest in the NLO materials that are already known; some of the most important will be described in Sect. 1.3. Finally, we will close this chapter with a very brief section (Sect. 1.4) that will provide the reader with useful conversion factors for the several units commonly used to report hyperpolarizabilities.

1.2 Some Important Nonlinear Optical Processes

As mentioned in the Introduction, parametric NLO phenomena may be either second- or third-order. Second-order processes occur only in crystals that lack inversion symmetry, such as lithium niobate or gallium arsenide, and in other noncentrosymmetric media, such as electric-field poled polymers or glasses. Since they are second-order they do not require fields as intense as those needed to generate observable third-order processes. On the other hand, third-order processes are more varied in nature due to the presence of the third field. The next subsection contains brief descriptions for some of the most important second-order processes including SHG, sum frequency generation, dc-Pockels effect, and optical rectification. This is followed by a second section concerning some of the most important third-order processes.

A compact summary is provided in the Table 1.1 below, which also provides a sample of applications.

1.2.1 Second-Order NLO Processes

Second-harmonic generation (SHG) is the prototype second-order NLO process in which two photons in a monochromatic light wave of frequency ω are effectively combined (or 'mixed') to form a single photon with double the frequency as shown

Table 1.1 Summary of some important second- and third-order nonlinear optical processes with sample applications

	Effects	Applications
$\chi^{(1)}$	Refractive index	Optical fiber
$\chi^{(2)}$	Second-harmonic generation ($\omega + \omega \rightarrow 2\omega$)	Frequency doubler
	Optical rectification ($\omega - \omega \rightarrow 0$)	Terahertz radiation
	Sum frequency generation ($\omega_1 + \omega_2 \rightarrow \omega_\sigma$)	IR surface spectroscopy
	Difference frequency generation ($\omega_1 - \omega_2 \rightarrow \omega_\sigma$)	Mid-range IR spectroscopy
	Pockels effect ($\omega + 0 \rightarrow \omega$)	Electro-optic modulator
$\chi^{(3)}$	Third-harmonic generation ($\omega + \omega + \omega \rightarrow 3\omega$)	Frequency tripler
	dc-SHG ($\omega + \omega + 0 \rightarrow 2\omega$)	Measurement of β
	Kerr effect ($\omega + 0 + 0 \rightarrow \omega$)	Ultra-fast optical switch
	Degenerate four-wave mixing ($\omega + \omega - \omega \rightarrow \omega$)	Holography, optical memory
	Optical mixing ($\omega_1 + \omega_2 + \omega_3 \rightarrow \omega_4$)	Raman spectroscopy
	EFIOR ($\omega - \omega + 0 \rightarrow 0$)	Filed induced properties

in Fig. 1.1 and Eq. (1.2). Famously, when the paper by Franken et al. demonstrating this effect appeared in Physical Review Letters [2], the dim spot at 2ω (wavelength $= 347.15$ nm) on the photographic paper could not be seen because the copy editor had mistaken it for a speck of dirt and removed it from the publication [5].

SHG is often employed to convert the output from an ultraviolet laser to a visible wavelength. Thus, the 1,064 nm output from a Nd:YAG laser can be fed through a KDP crystal (see later) to generate 532 nm green light. Similarly, the 808 nm output from a Ti:sapphire laser can be converted to 404 nm violet light. In biological and medical science, SHG is utilized for high-resolution optical microscopy. Non-centrosymmetric biological materials, such as collagen (found in most load-bearing tissues), will act as an SHG medium. By using an appropriate set of filters the incident light from a short-pulse femtosecond laser, can be easily separated from the emitted, frequency-doubled, SHG signal. This allows for very high axial and lateral resolution, comparable to that of confocal microscopy, but without having to use pinholes.

Sum frequency generation (SFG) is sometimes utilized generically to describe any NLO process where two input frequencies (ω_1, ω_2) are combined to form an outgoing light wave of frequency $\omega_\sigma = \omega_1 + \omega_2$. In that sense, SHG can be viewed as a special case of SFG. The latter is also used specifically for the case where one of the two input frequencies is in the visible range of the spectrum while the other lies in the infrared (IR) region. By means of this process in situ vibrational spectroscopy at interfaces can be carried out by employing a tunable IR laser.

The dc-pockels effect (dc-P), also known as the **electro-optic pockels effect (EOPE)** was discovered in 1893 by Friedrich Carl Alwin Pockels (1865–1913) when he found that a static dc electric field applied to certain birefringent materials causes the refractive index to vary, approximately in proportion to the strength of the field.

The coefficient of proportionality is $\chi^{(2)}(-\omega; \omega, 0)$ and this effect is the basis for Pockels cells, which have many applications. When used in conjunction with polarizers these cells can function either as on-off switches or as modulators for the optical rotation. They are also frequently employed in laser cavities as Q-switches that allow light to exit in short, intense output pulses. Pockels cells have additional applications in many different areas including quantum cryptography, electro-optic probes and videodisc mastering.

Optical rectification (OR), also known as **electro-optic rectification (EOR)** is an example of difference frequency mixing. In this case two input waves of frequency ω_1 and ω_2 mix to form an outgoing wave of frequency $\omega_\sigma = \omega_1 - \omega_2$. If $\omega_1 = \omega_2$, then a static DC-field is produced proportional to $\chi^{(2)}(0; \omega, -\omega)$. In practice, instead of total frequency cancellation, there is only quasi-cancellation because the input from, say, a femtosecond pulsed laser will contain a band of frequencies. The resulting emission, then, occurs in the terahertz (THz) region of the spectrum. Indeed, this is one of the main mechanisms for generating THz radiation using lasers. The EOR can be enhanced by introducing a metal surface.

1.2.2 Third-Order NLO Processes

In general the third-order susceptibilities may be written as $\chi^{(3)}(-\omega_\sigma; \omega_1, \omega_2, \omega_3)$, where the frequency of the induced dipole is $\omega_\sigma = \omega_1 + \omega_2 + \omega_3$. Such processes are observed for both centrosymmetric and non-centrosymmetric NLO materials. They require larger fields than second-order processes, but there are also more possibilities. Some of the most important correspond to the situation where the incident radiation is provided by a monochromatic laser of frequency ω either with or without the presence of a static DC field.

In the absence of a static DC field the input frequency can be tripled through **third harmonic generation (THG)** which corresponds to the case of optical mixing with $\omega_1 = \omega_2 = \omega_3 = \omega$. THG is commonly used to produce high intensity laser-driven nuclear fusion and, in some instances, to identify the interface between materials of different susceptibility.

A second process that occurs in the absence of a DC field is **degenerate four-wave mixing (DFWM)**. This corresponds to the case $\omega_1 = \omega_2 = \omega$, $\omega_3 = -\omega$ and is also known as the **intensity-dependent refractive index (IDRI)**. DFWM is used to achieve optical phase conjugation, which yields applications in holography, nonlinear microscopy, and in the characterization of nonlinear optical materials. The fact that the index of refraction is intensity-dependent accounts for utilization of DFWM for optical memory elements as well as optical computing. A related process is Coherent anti-Stokes Raman spectroscopy (CARS). In that case $\omega_3 = -\omega_\sigma$ with $\omega - \omega_\sigma$ equal to a Raman-active vibrational frequency.

The presence of a static DC field can lead to electric field induced properties akin to the second-order processes that arise in the absence of a DC field. For example, one can have **electric field induced second harmonic generation (EFISHG)** =

DC-SHG or **electric field induced optical rectification (EFIOR)**. The former is characterized by ($\omega_1 = \omega_2 = \omega$, $\omega_3 = 0$) and the latter by ($\omega_1 = \omega$, $\omega_2 = -\omega$, $\omega_3 = 0$). In non-centrosymmetric media the corresponding second- and third-order processes occur simultaneously and the third-order process may be largely masked.

Finally, the **electro-optical kerr effect (EOKE)** is akin to the DC-P effect except that it is quadratic in the static DC field rather than linear and it occurs also in centrosymmetric media. That is to say, EOKE refers to the case where ($\omega_1 = \omega$, $\omega_2 = \omega_3 = 0$). Again, the static field causes the sample to become birefringent with the difference in refractive index between light polarized in the parallel versus perpendicular directions now being directly proportional to the square of the DC field strength (and to the wavelength). This process is used, as is the DC-Pockels effect, to modulate the transmission of light by turning the DC field on and off. Since the response time is quite fast very rapid modulation can be achieved. The EOKE is employed, for example, in shutters for high speed photography.

There is also an AC, or simply, *optical kerr effect*. In that case, the incident light itself provides the electric field intensity (i.e. it replaces the external DC field). This effect governs several important processes associated with ultra-short intense laser pulses such as self-focusing and self-phase modulation.

1.2.3 Other NLO Processes

The previous two subsections are not meant to be exhaustive, but merely to whet the reader's appetite. There are a number of books that can be consulted for a more comprehensive listing, see, for example books as listed in [6–11]. From a computational perspective, the key point is that to take account of any second-order process it is sufficient to be able to calculate $\chi^{(2)}(-\omega_\sigma; \omega_1, \omega_2)$ with ω_1, ω_2 equal to arbitrary (positive or negative) frequencies and $\omega_\sigma = \omega_1 + \omega_2$. Likewise, for all third-order processes one needs $\chi^{(3)}(-\omega_\sigma; \omega_1, \omega_2, \omega_3)$ with arbitrary $\omega_1, \omega_2, \omega_3$ and $\omega_\sigma = \omega_1 + \omega_2 + \omega_3$.

There are other nonlinear optical processes involving magnetic as well as electric fields, such as the magnetic optical Kerr effect and Faraday rotation. They are certainly of considerable interest, but are beyond the scope of this monograph.

1.3 Nonlinear Optical Materials

Although a variety of NLO materials have found practical application, the search for new and improved materials continues briskly. One obvious general criterion is that the susceptibility for the particular frequency-dependent process involved should be large, whereas any interfering processes should be weak. In addition, the material should be transparent to both the impinging laser light and the modified beam while, for high efficiency, the phase matching condition must be well-satisfied. Finally, the sensitivity to laser damage as well as temperature fluctuations should be minimal along with favorable cost and availability. Of course, there will often be other criteria that depend upon the specific application.

Early on (and still today) the most commonly utilized NLO materials were inorganic crystals such as BBO (β-barium borate), KDP (potassium dihydrogen phosphate), KTP (potassium titanyl phosphate), and lithium niobate. These crystals are strongly birefringent, which is preferred for phase matching and, to a greater or lesser extent, meet the criteria mentioned above. On the other hand, organic NLO materials are on the ascendancy because they are easily modified chemically, can often be made cheaply, and high efficiencies are potentially achievable. Moreover, they occur in a wide variety of chemical systems including conjugated polymers, fullerenes, bi-/multi-radicals, electrides, graphenes, and others. Clearly, there exist numerous possibilities going forward.

1.4 Conversion Factors

NLO is a field that spans physics, chemistry and materials. Since different units are used by practitioners in each of these areas, as well as by experimentalists and theoreticians within the same area, it is important to be able to convert readily from one unit to another. For that purpose Table 1.2 lists the conversion factors between atomic units (a.u.), SI units, and electrostatic units (esu) for first- and second-order electric hyperpolarizabilities (β and γ) and, to be complete, for the electric dipole moment and linear polarizability (μ and α) as well [12].

Other useful conversion factors are:

Dipole moment μ: $1\,\mathrm{D}$ (Debye) $= 10^{-18}\,\mathrm{esu} = 3.33557 \times 10^{-30}\,\mathrm{C\,m}$.
Macroscopic second-order nonlinear optical susceptibility d_{33}: $1\,\mathrm{pm/V} = 2.387 \times 10^9\,\mathrm{esu}$.

Table 1.2 Conversion factors between atomic unit (a.u.), SI units, and electrostatic units (esu) for electric properties

	a.u.	SI	esu
μ	$1ea_0$	$8.478358 \times 10^{-30}\,\mathrm{Cm}$	2.5418×10^{-18}
α	$1a_0^3$	$1.648778 \times 10^{-41}\,\mathrm{C^2\,m^2\,J^{-1}}$	1.4817×10^{-25}
β	$1a_0^5/e$	$3.206361 \times 10^{-53}\,\mathrm{C^3\,m^3\,J^{-2}}$	8.6392×10^{-33}
γ	$1a_0^7/e^2$	$6.235377 \times 10^{-65}\,\mathrm{C^4\,m^4\,J^{-3}}$	5.0367×10^{-40}

References

1. Maiman, T.H.: Optical and microwave-optical experiments in ruby. Phys. Rev. Lett. **4**, 564–566 (1960)
2. Franken, P., Hill, A., Peters, C., Weinreich, G.: Generation of optical harmonics. Phys. Rev. Lett. **7**, 118–119 (1961)
3. Armstrong, J.A., Bloembergen, N., Ducuing, J., Pershan, P.S.: Interactions between light waves in a nonlinear dielectric. Phys. Rev. **127**, 1918–1939 (1962)
4. Bloembergen, N., Pershan, P.S.: Light waves at boundary of nonlinear media. Phys. Rev. **128**, 606–622 (1962)
5. Haroche, S.: Essay: fifty years of atomic, molecular and optical physics in physical review letters. Phys. Rev. Lett. **101**, 160001 (2008)
6. Sauter, E.G.: Nonlinear Optics. Wiley, New York (1996)
7. He, G., Liu, S.H.: Physics of Nonlinear Optics. World Scientific, Singapore (1999)
8. Agrawal, G.: Nonlinear Fiber Optics. Academic Press, Boston (2001)
9. Shen, Y.R.: The Principles of Nonlinear Optics. Wiley-Interscience, New York (2002)
10. Boyd, R.W.: Nonlinear Optics. Academic Press, Orlando (2008)
11. Agrawal, G.: Applications of Nonlinear Fiber Optics. Academic Press, New York (2010)
12. Shelton, D.P., Rice, J.E.: Measurements and calculations of the hyperpolarizabilities of atoms and small molecules in the gas phase. Chem. Rev. **94**, 3–29 (1994)

Chapter 2
Quantum-Mechanical Treatment of Responses to Electric Fields—Molecular Systems

Abstract In this chapter we first give a brief overview of theoretical methods for calculating the responses of smaller or larger, finite systems to electric fields. Subsequently, we concentrate on the quantum-mechanical (coupled) perturbation theory treatment of these systems. Both electronic and vibrational responses are discussed.

2.1 Introduction

In parallel with the many experimental studies of linear and non-linear optical properties, some of which were mentioned in the previous chapter, there has also been much theoretical work. We do not intend to describe all that has been done (for a relatively recent review, see [1]) but will present in this chapter an overview of the quantum-mechanical methods used to treat the electronic and nuclear responses to applied electric fields. These responses determine the linear and nonlinear optical (L&NLO) properties. Although the methodology has been developed in connection with ordinary small and medium-size molecules, our emphasis will be on treatments that are most readily extended to large systems, since the latter are of primary interest in this monograph.

The response of any system to applied electric fields, static and/or dynamic, can be calculated by solving the time-dependent Schrödinger equation

$$\left[\hat{H}_0(\mathbf{X}, \mathbf{x}) + \hat{H}'(\mathbf{X}, \mathbf{x}, t) \right] \Psi(\mathbf{X}, \mathbf{x}, t) = i\hbar \frac{\partial}{\partial t} \Psi(\mathbf{X}, \mathbf{x}, t). \qquad (2.1)$$

Here $\hat{H}_0$ is the Hamilton operator for the field-free system, whereas the effect of the field(s) is described through $\hat{H}'$ (Sect. 2.3 and Chap. 3). The L&NLO properties of concern here are defined in terms of a power series expansion for the response to the latter. In Eq. (2.1) $\mathbf{X}$ denotes the set of nuclear coordinates, whereas $\mathbf{x}$ serves the same role for the electronic coordinates. We have assumed that $\hat{H}_0$ does not contain any explicit time dependence and, for sake of simplicity, relativistic effects are ignored (i.e. there are no terms in the Hamiltonian involving electron spin operators).

Within the Born-Oppenheimer approximation, the solutions of the field-free time-independent Schrödinger equation may be written as products of electronic and

© The Author(s) 2015
F.L. Gu et al., *Calculations on nonlinear optical properties for large systems*,
SpringerBriefs in Electrical and Magnetic Properties of Atoms,
Molecules, and Clusters, DOI 10.1007/978-3-319-11068-4_2

vibrational, i.e. vibronic, wavefunctions where the electronic wavefunction satisfies

$$\hat{H}_e^0(\mathbf{X}; \mathbf{x})\Phi_K^0(\mathbf{X}; \mathbf{x}) = E_K^0(\mathbf{X})\Phi_K^0(\mathbf{X}; \mathbf{x}). \tag{2.2}$$

In the above equation the electronic Hamiltonian, $\hat{H}_e^0(\mathbf{X}; \mathbf{x})$, is the same as $\hat{H}_0$ except that the nuclear kinetic energy term has been removed. Besides the electronic kinetic energy this Hamiltonian contains the electrostatic potential for interaction between the electrons and nuclei as well as the electron-electron and nuclear-nuclear repulsion, all evaluated at a fixed geometry. Thus, $E_K^0(\mathbf{X})$ and $\Phi_K^0(\mathbf{X}; \mathbf{x})$ depend parametrically on the positions of the nuclei. Using the electronic wavefunctions of Eq. (2.2) as the basis for a time-dependent perturbation treatment of $\hat{H}'(\mathbf{X}, \mathbf{x}, t)$ one can obtain the conventional clamped nucleus (CN) sum-over-state (SOS) expressions presented in Sect. 2.3 for the time- or frequency-dependent electronic (hyper)polarizabilities at, say, the ground state equilibrium geometry. These expressions are approximate, even when zero point vibrational averaging is included, because the effect of non-adiabatic coupling with simultaneous vibrational motions is ignored [2]. Nonetheless, in the cases that have been studied the CN approximation has proved to be quite accurate and, almost without exception, is used in the SOS treatment of electronic L&NLO properties as it is here.

SOS calculations require determining the entire set of excited electronic states, which is impractical for ab initio treatment of large (or, even, medium-size) systems. Such calculations have been widely employed, however, at the semi-empirical level [3], and/or in cases where it is assumed that only a few key states are important (e.g. 2-state approximation for β), or when approximate ab initio excited states are taken to be sufficient (e.g. in the uncoupled Hartree-Fock method discussed briefly in Sect. 2.5). As a result, SOS expressions can be useful for qualitative analysis. They have also proved valuable for other purposes as mentioned in Sect. 2.2.

At the ab initio level, for large systems it is preferable to employ either the time-dependent Hartree-Fock (TDHF) or time-dependent Kohn-Sham DFT (TDDFT) procedure. In Sect. 2.3 the general formulation of these two analytical single-particle methods is outlined. Then, in Sect. 2.4, we present the resulting perturbation theory equations in compact form and describe the strategy for solving them. In passing we note that, beyond TDHF, any correlated wavefunction method (MP2, MCSCF, CCSD, ...) can be adapted for calculating electronic L&NLO properties, although these procedures are much more readily applicable to smaller systems.

The electronic response to a static external field is a special case. In that event the clamped nucleus electronic (hyper)polarizabilities can alternatively be obtained numerically by means of the finite field (FF) method. In the FF method the perturbation term $[\hat{H}'(\mathbf{X}, \mathbf{x}, 0)$ in Eq. (2.1)] is included directly in $\hat{H}_e$. Calculations are carried out for different field magnitudes, as well as directions, and then, the electronic response is fit to a power series in the magnitude for each desired direction. In principle, the FF approach can be extended to the time domain to find dynamic electronic (hyper)polarizabilities as well.

As implied above, there is a separate nuclear (i.e. vibrational) response that contributes to the (hyper)polarizabilities. It is the subject of Sect. 2.5. This contribution appears in the theory when one takes into account the vibrational component of the Born-Oppenheimer vibronic wavefunction. Then, it is readily seen that the sum-over- (vibronic) states contains terms due to vibrational excitations on the ground electronic state potential energy surface [4]. In contrast with the small error in the electronic (hyper)polarizabilities due to the CN approximation, the effect of these vibrational response terms can be important—sometimes much more so than the CN electronic terms—as will be seen in Sect. 2.5, where the role of FF calculations in determining this effect will be elucidated.

2.2 Clamped Nucleus Sum-Over-States Electronic (Hyper)polarizabilities

In order to obtain an expression for the electronic (hyper)polarizabilities we clamp the nuclear coordinates at $\mathbf{X} = \mathbf{X}_0$. Then, in the field-free case, the time-dependent solutions of the electronic Schrödinger equation may be written as

$$\Phi_K^0(\mathbf{X}_0; \mathbf{x}, t) = \exp(-i E_K^0 t/\hbar) \Phi_K^0(\mathbf{X}_0; \mathbf{x}) \tag{2.3}$$

where $\Phi_K^0(\mathbf{X}_0, \mathbf{x})$ and $E_K^0 = E_K^0(\mathbf{X}_0)$ are the stationary state solutions of Eq. (2.2). For spatially uniform oscillating electric fields ($\omega = 0$ in static limit),

$$\mathbf{E}(t) = \sum_\omega [\mathbf{E}_{+\omega} e^{+i\omega t} + \mathbf{E}_{-\omega} e^{-i\omega t}], \quad \text{with } \mathbf{E}_{+\omega} = \mathbf{E}_{-\omega}, \tag{2.4}$$

the scalar interaction with the molecular dipole moment operator ($-e$ is the electronic charge and $e Z_A$ the charge of nucleus A)

$$\hat{\mu}(\mathbf{X}_0, \mathbf{x}) = -e \sum_i \mathbf{x}_i + e \sum_A Z_A \mathbf{X}_{A,0} \tag{2.5}$$

is given by:

$$\hat{H}'(\mathbf{X}_0, \mathbf{x}, t) = -\hat{\mu}(\mathbf{X}_0, \mathbf{x}) \cdot \mathbf{E}(t). \tag{2.6}$$

As shown by Orr and Ward [5], a standard time-dependent perturbation theory treatment of Eqs. (2.1), (2.3) and (2.6), assuming resonance-induced excited state populations and damping are negligible, leads to the following sum-over-states (SOS) formulas for the dynamic (i.e. frequency-dependent) (hyper)polarizabilities

$$\alpha_{\zeta\eta}(-\omega_\sigma; \omega) = \frac{1}{\hbar} \sum P_{-\sigma,1} \sum_K{}' \frac{1}{\omega_K - \omega_\sigma} \langle 0|\hat{\mu}_\zeta|K\rangle\langle K|\hat{\mu}_\eta|0\rangle \tag{2.7}$$

$$\beta_{\zeta\eta\kappa}(-\omega_\sigma;\omega_1,\omega_2) = \frac{1}{\hbar^2}\sum P_{-\sigma,1,2}\sideset{}{'}\sum_K\sideset{}{'}\sum_L\frac{1}{(\omega_K-\omega_\sigma)(\omega_L-\omega_2)}$$

$$\times\langle 0|\hat{\mu}_\zeta|K\rangle\langle K|\bar{\hat{\mu}}_\eta|L\rangle\langle L|\hat{\mu}_\kappa|0\rangle \tag{2.8}$$

$$\gamma_{\zeta\eta\kappa\lambda}(-\omega_\sigma;\omega_1,\omega_2,\omega_3) = \frac{1}{\hbar^3}\sum P_{-\sigma,1,2,3}$$

$$\left[\sideset{}{'}\sum_K\sideset{}{'}\sum_L\sideset{}{'}\sum_M\frac{1}{(\omega_K-\omega_\sigma)(\omega_L-\omega_2-\omega_3)(\omega_M-\omega_3)}\right.$$

$$\times\langle 0|\hat{\mu}_\zeta|K\rangle\langle K|\bar{\hat{\mu}}_\eta|L\rangle\langle L|\bar{\hat{\mu}}_\kappa|M\rangle\langle M|\hat{\mu}_\delta|0\rangle$$

$$-\sideset{}{'}\sum_K\sideset{}{'}\sum_L\frac{1}{(\omega_K-\omega_\sigma)(\omega_L-\omega_3)(\omega_L+\omega_2)}$$

$$\left.\times\langle 0|\hat{\mu}_\zeta|K\rangle\langle K|\hat{\mu}_\eta|0\rangle\langle 0|\hat{\mu}_\kappa|L\rangle\langle L|\hat{\mu}_\lambda|0\rangle\right]. \tag{2.9}$$

In Eqs. (2.7), (2.8) and (2.9) α is the linear polarizability, β is the first hyperpolarizability, and γ is the second hyperpolarizability. The symbol $\beta_{\zeta\eta\kappa}(-\omega_\sigma;\omega_1,\omega_2)$, for example, indicates the $\zeta\eta\kappa$ tensor component of the first hyperpolarizability for applied fields of frequency ω_1 (in the Cartesian direction η) and ω_2 (in the direction κ); $\omega_\sigma = \omega_1 + \omega_2$ is the frequency of the induced dipole moment (in the direction ζ); and $\sum P_{-\sigma,1,2}$ represents a sum over the 6 permutations of the pairs ($-\omega_\sigma/\hat{\mu}_\zeta$, $\omega_1/\hat{\mu}_\eta$, and $\omega_2/\hat{\mu}_\kappa$). We use primes on the sums over K, L to indicate that the ground electronic state $|0\rangle$ is excluded. The quantity $\hbar\omega_K$ is the energy of electronic state $|K\rangle$ relative to $|0\rangle$, and $\bar{\hat{\mu}} = \hat{\mu} - \langle 0|\hat{\mu}|0\rangle$. An exactly analogous interpretation applies to Eqs. (2.7) and (2.9).

In general, the SOS expressions are too inefficient computationally for quantitative purposes, but they can be useful for qualitative analysis as mentioned in the overview preceding this section. Moreover, they often serve as the basis for the formulation of: (i) resonant NLO processes such as two-photon absorption [$\omega_2 = \omega_3 = \omega = \omega_\sigma$ and $\omega_L = 2\omega$ in the first term of Eq. (2.9)]; (ii) vibrational (hyper)polarizabilities (see Sect. 2.5); and (iii) physical limits on off-resonance electronic (hyper) polarizabilities [6, 7]. Finally, if $|0\rangle$ is the Hartree-Fock (HF) wavefunction, then the SOS formulas become equivalent to an uncoupled time-dependent Hartree-Fock (UC-TDHF) perturbation treatment.

2.3 Time-Dependent Hartree-Fock and DFT (Clamped Nucleus) Electronic Properties

The UC-TDHF perturbation method mentioned in the previous section does not account for orbital relaxation, i.e. the change in the HF density matrix (see below) induced by the applied fields. Such orbital relaxation is included in the (coupled) TDHF treatment and, typically, makes an important contribution to the calculated electronic linear and nonlinear (L&NLO) optical properties. In the limit of static fields the TDHF method is often referred to as coupled perturbed Hartree Fock (CPHF).

There are a number of alternative analytical procedures for carrying out a TDHF, also known as the Random Phase Approximation (RPA), treatment (see, e.g. [8–10]). In fact, TDHF = RPA is the simplest version of the (more general) 'response method', often developed using polarization propagators [11, 12]. For the treatment of large systems this simplest (i.e. TDHF) version is the most useful. One convenient formulation of TDHF in the non-resonant regime is the following procedure due to Karna and Dupuis [9] (see references cited therein for earlier work). Using a single determinant wavefunction for $|K\rangle = |0\rangle$ in Eq. (2.2), and variationally optimizing the HF orbitals ($\psi_i^{(0)}$), leads to the usual field-free Fock equation (in matrix form)

$$\underline{\underline{F}}^{(0)}\underline{\underline{C}}^{(0)} = \underline{\underline{S}}^{(0)}\underline{\underline{C}}^{(0)}\underline{\underline{\varepsilon}}^{(0)}, \quad \left(\underline{\underline{C}}^{(0)\dagger}\underline{\underline{S}}^{(0)}\underline{\underline{C}}^{(0)} = \underline{\underline{1}}\right) \tag{2.10}$$

where $\underline{\underline{C}}^{(0)}$ is the matrix of expansion coefficients that transform the basis functions $\underline{\chi}$ into molecular orbitals $\underline{\psi}^{(0)}$,

$$\psi_i^{(0)}(\mathbf{X}_0, \mathbf{x}) = \sum_j \chi_j(\mathbf{X}_0, \mathbf{x})C_{ji}^{(0)}(\mathbf{X}_0) \tag{2.11}$$

or

$$\underline{\psi}^{(0)} = \underline{\chi}\,\underline{\underline{C}}^{(0)}. \tag{2.12}$$

In Eq. (2.10) $\underline{\underline{\varepsilon}}^{(0)}$ is normally (= the canonical choice) taken to be a diagonal matrix of Lagrange multipliers—also known as orbital energies). Moreover, $\underline{\underline{S}}^{(0)}$ is the overlap matrix,

$$S_{ij}^{(0)} = \langle\chi_i|\chi_j\rangle \tag{2.13}$$

and, assuming that all occupied molecular orbitals are doubly occupied,

$$F_{ij}^{(0)} = \langle\chi_i(1)|\hat{h}(1)|\chi_j(1)\rangle$$
$$+ \sum_{kl}\left[\langle\chi_i(1)\chi_k(2)|\frac{1}{r_{12}}|\chi_j(1)\chi_l(2)\rangle - \frac{1}{2}\langle\chi_i(1)\chi_k(2)|\frac{1}{r_{12}}|\chi_l(1)\chi_j(2)\rangle\right]D_{kl}^{(0)}. \tag{2.14}$$

The operator $\hat{h}(1)$ here contains the one-electron kinetic energy and nuclear-electron attraction terms while the remaining terms on the rhs, due to electron-electron repulsion, consist of two-electron integrals each multiplied by an element of the density matrix

$$\underline{\underline{D}}^{(0)} = \underline{\underline{C}}^{(0)} \underline{\underline{n}}\, \underline{\underline{C}}^{(0)\dagger} \tag{2.15}$$

where $\underline{\underline{n}}$ is the diagonal occupation matrix containing the eigenvalues 2 for occupied orbitals and 0 for unoccupied orbitals. It is straightforward to extend the above equations to unrestricted Hartree-Fock with singly-occupied spin-orbitals.

When the perturbation due to the terms in Eq. (2.6) is taken into account the time-dependent Schrödinger equation must be utilized. Thus, $i\,\underline{\underline{S}}^{(0)} \frac{\partial}{\partial t}\underline{\underline{C}}$ must be added to the rhs of Eq. (2.10) and the perturbed coefficient matrix (which depends upon the frequency and direction of the field), as well as the corresponding matrix of Lagrange multipliers, becomes time-dependent. At the same time, it is convenient to retain the normalization condition so that $(S = S^{(0)})$:

$$\underline{\underline{C}}^{\dagger} \underline{\underline{S}}\, \underline{\underline{C}} = \underline{\underline{C}}^{(0)\dagger} \underline{\underline{S}}\, \underline{\underline{C}}^{(0)} = \underline{\underline{1}}. \tag{2.16}$$

Subsequently, we may expand all field-dependent quantities in the Fock equation (2.10) (the basis functions χ are assumed here, and above, to be field-independent) as power series in $E_\omega e^{\pm i\omega t}$. Then, terms on either side of this equation, that are of like power in the field and have the same exponential frequency factor, are equated to one another. This leads to the TDHF perturbation equations, which will be presented in a compact form in the next section. The general strategy for solving them is described below with more details given later.

For non-resonant frequencies well below the first electronic absorption the first-order TDHF perturbation equation for the coefficient matrix may be solved self-consistently starting with the uncoupled approximation. This matrix (as well as the matrix of Lagrange multipliers) is determined only up to an arbitrary unitary trans-formation amongst the occupied (and/or amongst unoccupied) molecular orbitals. In order to ensure stable solutions a non-canonical choice is made whereby $\underline{\underline{\varepsilon}}$ has non-zero off-diagonal elements in first-order connecting different occupied (as well as different unoccupied) molecular orbitals, while maintaining the fact that there are no elements connecting the block of occupied orbitals with the block of unoccupied orbitals. The non-zero off-diagonal elements of the Lagrange multiplier matrix within the occupied and unoccupied blocks are determined by the orthonormality condition for the first-order coefficients, which is enforced in a particularly simple manner (see Sect. 2.4). The second-order perturbation equations may be solved similarly using the first-order solutions and a computationally convenient non-canonical choice for the second-order Lagrange multipliers (similar to what is done in first-order). Explicit expressions for the non-canonical perturbation treatment are provided in the next section.

Finally, the total (permanent + field-induced) dipole moment is just the average value of the dipole moment operator

$$\mu = \sum_{ij} \langle \chi_i | \hat{\mu} | \chi_j \rangle D_{ji} = \mathrm{Tr}(\underline{\underline{M}}\, \underline{\underline{D}}). \tag{2.17}$$

Here $\underline{\underline{D}}$ is the total density matrix. Hence, the terms in $\underline{\underline{D}}$ that are linear in the field determine the linear polarizability tensor; the quadratic terms determine the first hyperpolarizability tensor; etc. These are the $n + 1$ expressions for the (hyper)polarizabilities. As in the time-independent case, the TDHF perturbation theory equations can be manipulated to yield a $2n + 1$ rule whereby not only the linear polarizability, but also the first hyperpolarizability (a third-order property) can be obtained from solutions of the first-order perturbation equations. For the second (hyper)polarizability, however, the second-order (but not third-order) solutions must be known as well. The $2n + 1$ formulas for the NLO hyperpolarizabilities induced by a monochromatic applied field, with and without an additional static (i.e., DC) field may be found in Tables VII and VIII of [9]. These formulas may readily be extended to cover the general case when there is more than one laser source each operating at its own frequency.

The TDHF method is a wavefunction approach that, by definition, does not take account of electron correlation. Correlation can be introduced through any of the standard quantum-chemical methods that have been extended to take into account time-dependence. This includes Møller-Plesset (MP) perturbation theory and coupled cluster methods, as well as the multi-configuration self-consistent field and configuration interaction treatments. There are also higher-order linear response/polarization propagator methods. As noted above, TDHF corresponds to the simplest possible version, which might be called the first-order polarization propagator approximation. The second-order polarization propagator approximation (SOPPA) [11], which provides correlated results of MP2 quality, has been developed for hyperpolarizabilities as well as linear polarizabilities [12].

Even for medium-size systems correlated calculations can be quite tedious. Thus, one may want to utilize a short-cut, even though some accuracy is lost. Assuming it is feasible to obtain the static correlated property, $P^{\mathrm{corr}}(\text{static})$, an estimate of the dynamic (non-resonant) value can be made by scaling the static result according to

$$P^{\mathrm{corr}}(\text{dynamic}) \simeq \frac{P^{\mathrm{corr}}(\text{static})}{P^{\mathrm{TDHF}}(\text{static})} P^{\mathrm{TDHF}}(\text{dynamic}). \tag{2.18}$$

Another possibility, when correlated frequency-dependent values for just a single dynamic property are available, is to obtain an approximation for other dynamic properties using a power series expansion through fourth-order in the optical frequencies. Through that order just three parameters, at most, determine the frequency-dependence of *all* monochromatic NLO processes [13].

A time-dependent density functional theory (TDDFT) treatment, based on the Kohn-Sham method, may be carried out in a manner that is similar to TDHF. The major formal difference between the two lies in the replacement of the TDHF exchange contribution [second term in the double sum of Eq. (2.14)] by a term that involves the time-dependent exchange-correlation (XC) potential, V_{ij}^{XC}. The latter, in turn, depends upon the time-dependent density function, $\rho(x, t)$. The potential V^{XC} comes in many different variants. For sake of simplicity we consider here only generalized gradient approximations (GGAs). However, our formulation is readily

extended to meta-GGA and also to hybrid functionals using the TDHF expression
for 'exact' exchange in Eq. (2.10).

From Eq. (2.15)

$$\rho = 2\sum_{i=1}^{\text{occ}} \psi_i^* \psi_i = \sum_{kl} D_{kl} \chi_k^* \chi_l \tag{2.19}$$

and, for GGAs,

$$\nabla\rho = \sum_{kl} D_{kl} \nabla(\chi_k^* \chi_l) \tag{2.20}$$

where the superscript $^{(0)}$ on the density matrix has been dropped because perturbation
corrections are now included. Hence, the density matrix and the density function are
both time-dependent. In principle, V^{XC} should be obtained by taking the functional
derivative of the XC action, A^{XC}, with respect to $\rho(r, t)$ [14], but in practice the
adiabatic approximation is employed, in which case A^{XC} is replaced by the time-
independent XC energy, E^{XC}. If we write

$$E^{\text{XC}} = \int f^{\text{XC}}(\rho, |\nabla\rho|^2) d\mathbf{x} \tag{2.21}$$

then, in the absence of any fields, it follows that the matrix elements of the XC
potential in the basis set χ are given by

$$V_{ij}^{\text{XC}} = \int \left[\frac{\partial f^{\text{XC}}}{\partial \rho} \chi_i^* \chi_j + 2\frac{\partial f^{\text{XC}}}{\partial |\nabla\rho|^2} \nabla\rho \cdot \nabla(\chi_i^* \chi_j) \right] d\mathbf{x} \tag{2.22}$$

as Pople et al. have shown using integration by parts [15].

The next step is to expand the integrand as a power series in the field(s). Thus, in
first-order, applying the chain rule to the $\frac{\partial f^{\text{XC}}}{\partial \rho}$ term gives rise to two contributions one
of which is $\frac{\partial^2 f^{\text{XC}}}{\partial \rho^2} \frac{d\rho}{dE_{\pm\omega}}\Big|_{E_{\pm\omega}=0}$. According to the adiabatic approximation $\frac{\partial^2 f^{\text{XC}}}{\partial \rho^2}$ is
evaluated as if the field were static. The other contribution arises from the derivative
of $\frac{\partial f^{\text{XC}}}{\partial \rho}$ with respect to $|\nabla\rho|^2$ evaluated, again, in the adiabatic approximation. There
are likewise two first-order contributions that occur when the chain rule is applied to
the second term in Eq. (2.22) plus a contribution from the derivative with respect to
$E_{\pm\omega}$ of the factor $\nabla\rho(x, t)$ appearing in that term. Thus, there are a total of five terms
involving first and second derivatives of f^{XC} needed for the first-order perturbation
equation. An explicit expression for static fields, that may be easily generalized for
the time-dependent case, is given in Ref. [16]. In second-order one also needs third
derivatives and the expressions become much messier (see [17]). Nonetheless, an
automatic procedure for determining the required derivatives is available [18, 19].
Of course, the integration in Eq. (2.22) must be carried out numerically just as in
ordinary field-free DFT.

Besides the field-dependent XC potential one must also evaluate derivatives of the potential (with respect to ρ and $|\nabla\rho|^2$) in order to obtain the complete XC contribution to the first and second hyperpolarizability. However, the $2n + 1$ rule is maintained. Again, the required expressions have been reported for static fields [16, 17] and, as noted above, the higher-order functional derivatives of f^{XC} can be determined automatically.

We have focused here on the formal aspects of TDDFT. The efficiency of any particular implementation will depend upon the computational strategy employed, especially with regard to the method of integration (grid-based or not) and whether density fitting is used. The advantage of density fitting for the polarizability of large molecules in static fields has been demonstrated [20].

In addition to the perturbation theory procedure, there is also the possibility of calculating frequency-dependent electronic (hyper)polarizabilities numerically through DFT molecular dynamics simulations. At the present time this approach remains to be fully explored.

2.4 Solving the TDHF and TDDFT Equations

The previous section contains an overview of the time-dependent Hartree-Fock (TDHF) and time-dependent DFT [or more precisely, time-dependent Kohn-Sham (TDKS)] perturbation theory methods. In this section we present explicit expressions for the perturbation equations and their solutions, which in general terms is the same in either case. For that purpose it is convenient to use the compact notation illustrated below for the Fock Hamiltonian matrix:

$$
\begin{aligned}
\underline{\underline{F}} = \underline{\underline{F}}^{(0)} &+ \sum_{\zeta} \underline{\underline{F}}_{\zeta}^{(1)} E_{\zeta} + \sum_{\zeta\eta} \underline{\underline{F}}_{\zeta\eta}^{(2)} E_{\zeta} E_{\eta} + \sum_{\zeta\eta\kappa} \underline{\underline{F}}_{\zeta\eta\kappa}^{(3)} E_{\zeta} E_{\eta} E_{\kappa} + \cdots \\
&\to F + \sum_{\zeta} F_{\zeta} E_{\zeta} + \sum_{\zeta\eta} F_{\zeta\eta} E_{\zeta} E_{\eta} + \sum_{\zeta\eta\kappa} F_{\zeta\eta\kappa} E_{\zeta} E_{\eta} E_{\kappa} + \cdots \quad (2.23)
\end{aligned}
$$

Note that the double underline to indicate a matrix is now omitted and the number of indices for the Cartesian directions ζ, η, ... gives the order of perturbation theory. The field-free matrix F is given by Eq. (2.14); F_{ζ} is obtained from the same equation by replacing $\hat{h}(1)$ with $-\hat{\mu}_{\zeta}(1)$ and $D^{(0)}$ with D_{ζ}; and finally, $F_{\zeta\eta}$, $F_{\zeta\eta\kappa}$, etc. are obtained for TDHF by deleting the $\hat{h}^{(1)}$ term in Eq. (2.14) and replacing $D^{(0)}$ with $D_{\zeta\eta}$, $D_{\zeta\eta\kappa}$, etc. For TDKS the Fock matrices are different, but can be determined as described in the previous section. The field frequencies are not shown in Eq. (2.23). They are, however, included later on. As before, we assume that the molecular orbitals are restricted to double occupancy so that there is no spin polarization. Finally, for a more detailed presentation of this notation and the following discussion we recommend Ref. [9].

Table 2.1 TDHF/TDKS perturbation equations and normalization conditions

Order	Perturbation equations	Normalization conditions
0th	$FC = SC\varepsilon$	$C^\dagger SC = 1$
1st	$F_\zeta(\omega_1)C + FC_\zeta(\omega_1) + \omega_1 SC_\zeta(\omega_1)$ $= SC_\zeta(\omega_1)\varepsilon + SC\varepsilon_\zeta(\omega_1)$	$C^\dagger SC_\zeta(\omega_1) + C_\zeta^\dagger(-\omega_1)SC = 0$
2nd	$F_{\zeta\eta}(\omega_1, \omega_2)C + F_\zeta(\omega_1)C_\eta(\omega_2) + F_\eta(\omega_2)C_\zeta(\omega_1)$ $+ FC_{\zeta\eta}(\omega_1, \omega_2) + (\omega_1 + \omega_2)SC_{\zeta\eta}(\omega_1, \omega_2)$ $= SC_{\zeta\eta}(\omega_1, \omega_2)\varepsilon + SC_\zeta(\omega_1)\varepsilon_\eta(\omega_2)$ $+ SC_\eta(\omega_2)\varepsilon_\zeta(\omega_1) + SC\varepsilon_{\zeta\eta}(\omega_1, \omega_2)$	$C^\dagger SC_{\zeta\eta}(\omega_1, \omega_2)$ $+ C_\eta^\dagger(-\omega_2)SC_\zeta(\omega_1)$ $+ C_{\zeta\eta}^\dagger(-\omega_1, -\omega_2)SC$ $+ C_\zeta^\dagger(-\omega_1)SC_\eta(\omega_2) = 0$

Table 2.1 contains the TDHF/TDKS perturbation equations and normalization conditions through second-order, which is sufficient to calculate α, β and γ according to the $2n+1$ rule. For many (but not all) NLO processes of interest ω_1 and ω_2 are equal to $\pm\omega$ $(\omega > 0)$ or 0. In the same notation, the corresponding density matrices are

0th-order:

$$D = CnC^\dagger$$

1st-order:

$$D_\zeta(\omega_1) = C_\zeta(\omega_1)nC^\dagger + CnC_\zeta^\dagger(-\omega_1)$$

and 2nd-order:

$$D_{\zeta\eta}(\omega_1, \omega_2) = C_{\zeta\eta}(\omega_1, \omega_2)nC^\dagger + C_\eta(\omega_2)nC_\zeta^\dagger(-\omega_1)$$
$$+ C_\zeta(\omega_1)nC_\eta^\dagger(-\omega_2) + CnC_{\zeta\eta}^\dagger(-\omega_1, -\omega_2). \qquad (2.24)$$

These density matrices are, as usual, in the atomic orbital representation.

As shown by Karna and Dupuis [9], the perturbation theory equations can be solved by introducing a set of transformation matrices, U,

$$C_\zeta(\omega_1) = CU_\zeta(\omega_1)$$
$$C_{\zeta\eta}(\omega_1, \omega_2) = CU_{\zeta\eta}(\omega_1, \omega_2), \qquad (2.25)$$

as well as the G matrices,

$$G_\zeta(\omega_1) = C^\dagger F_\zeta(\omega_1)C$$
$$G_{\zeta\eta}(\omega_1, \omega_2) = C^\dagger F_{\zeta\eta}(\omega_1, \omega_2)C. \qquad (2.26)$$

which are both represented in the molecular orbital basis. Multiplying the perturbation equations from the left by $C^\dagger$, one obtains relations that immediately yield the

Table 2.2 TDHF/TDKS Lagrange multiplier matrices

Order	Equation
0th	$\varepsilon = C^\dagger F C$
1st	$\varepsilon_\zeta(\omega_1) = G_\zeta(\omega_1) + \varepsilon U_\zeta(\omega_1) - U_\zeta(\omega_1)\varepsilon + \omega_1 U_\zeta(\omega_1)$
2nd	$\varepsilon_{\zeta\eta}(\omega_1, \omega_2) = G_{\zeta\eta}(\omega_1, \omega_2) + G_\zeta(\omega_1)U_\eta(\omega_2) + G_\eta(\omega_2)U_\zeta(\omega_1)$
	$\quad + \varepsilon U_{\zeta\eta}(\omega_1, \omega_2) - U_{\zeta\eta}(\omega_1, \omega_2)\varepsilon$
	$\quad - U_\zeta(\omega_1)\varepsilon_\eta(\omega_2) - U_\eta(\omega_2)\varepsilon_\zeta(\omega_1) + (\omega_1 + \omega_2)U_{\zeta\eta}(\omega_1, \omega_2)$

Lagrange multiplier matrices reported in Table 2.2. Then, the off-diagonal blocks of the U matrices (that connect the set of occupied orbitals with the set of unoccupied orbitals) are determined by the fact that the corresponding blocks of the Lagrange multiplier matrices (see Table 2.2) must vanish. The diagonal blocks of the U matrices are determined by the normalization conditions. Substitution of Eq. (2.24) into these conditions gives

$$U_\zeta(\omega_1) + U_\zeta^\dagger(-\omega_1) = 0$$
$$U_{\zeta\eta}(\omega_1, \omega_2) + U_\zeta^\dagger(-\omega_1)U_\eta(\omega_2) + U_\eta^\dagger(-\omega_2)U_\zeta(\omega_1) + U_{\zeta\eta}^\dagger(-\omega_1, -\omega_2) = 0. \quad (2.27)$$

For the non-canonical solution one makes the choice

$$U_\zeta(\omega_1) = U_\zeta^\dagger(-\omega_1)$$
$$U_{\zeta\eta}(\omega_1, \omega_2) = U_{\zeta\eta}^\dagger(-\omega_1, -\omega_2) \quad (2.28)$$

which leads to the diagonal blocks of U in Table 2.3. The results shown are expressed in terms of T matrices also reported in the table.

From the solutions to the TDHF or TDKS equations one can directly obtain the $n + 1$ rule (hyper)polarizabilities for both static and dynamic fields. The polarizabilities are given by

$$\alpha_{\zeta\eta}(\mp\omega_1; \pm\omega_1) = -\mathrm{Tr}\left[M_\zeta D_\eta(\pm\omega_1)\right], \quad (2.29)$$

Table 2.3 Solutions for the TDHF/TDKS coefficient matrices in terms of U [see Eq. (2.25)]

Matrix	1st order	2nd order
T-matrix		$T_{\zeta\eta}(\omega_1, \omega_2) = G_\zeta(\omega_1)U_\eta(\omega_2)$
		$\quad - U_\zeta(\omega_1)G_\eta(\omega_2)$
		$\quad + G_\eta(\omega_2)U_\zeta(\omega_1)$
		$\quad - U_\eta(\omega_2)G_\zeta(\omega_1)$
U-matrix block diagonal elements	$U_\zeta(\omega_1) = 0$	$U_{\zeta\eta}(\omega_1, \omega_2) = \frac{1}{2} \times [U_\zeta(\omega_1)U_\eta(\omega_2)$
		$\quad + U_\eta(\omega_2)U_\zeta(\omega_1)]$
U-matrix non-diagonal blocks	$U_{\zeta,ij}(\omega_1)$	$U_{\zeta\eta,ij}(\omega_1, \omega_2)$
	$= \frac{G_{\zeta,ij}(\omega_1)}{\varepsilon_j - \varepsilon_i - \omega_1}$	$= \frac{G_{\zeta\eta,ij}(\omega_1,\omega_2) + T_{\zeta\eta,ij}(\omega_1,\omega_2)}{\varepsilon_j - \varepsilon_i - (\omega_1 + \omega_2)}$

where M_ζ is the ζ component of the dipole moment matrix in Eq. (2.17) and D_η is defined in Eq. (2.24). Similarly, the first hyperpolarizabilities can be expressed in terms of the second-order density matrices:

$$\beta_{\zeta\eta\kappa}(-\omega_1 - \omega_2; \omega_1, \omega_2) = -\text{Tr}\left[M_\zeta D_{\eta\kappa}(\omega_1, \omega_2)\right]. \qquad (2.30)$$

and the second hyperpolarizabilities in terms of the third-order density matrices

$$\gamma_{\zeta\eta\kappa\lambda}(-\omega_1 - \omega_2 - \omega_3; \omega_1, \omega_2, \omega_3) = -\text{Tr}\left[M_\zeta D_{\eta\kappa\lambda}(\omega_1, \omega_2, \omega_3)\right]. \qquad (2.31)$$

The conversion of Eqs. (2.30) and (2.31) into $2n + 1$ rule formulas requires a complicated sequence of steps that will not be presented here. A fairly compact general expression can be developed for the first hyperpolarizability. There are additional terms in TDKS that are not present in TDHF. These were discussed in the previous section where appropriate references were cited.

The analogous result for the second hyperpolarizability is much less compact. Explicit expressions for monochromatic processes, in the presence or absence of a static field may be found in Table VIII of Ref. [9]. Some of the more important second- and third-order NLO properties were described in Chap. 1 where a tabular summary (see Table 1.1) is also provided.

2.5　Vibrational Linear and Nonlinear Polarizabilities

The vibrational (hyper)polarizability is **not** the contribution due to zero-point vibrational averaging. That is a much smaller effect. In order to explain the origin of the vibrational hyperpolarizability we return to the SOS formulas for the electronic properties in Eqs. (2.7), (2.8) and (2.9). Those expressions were obtained using the pure electronic wavefunctions of Eq. (2.2) with the nuclei clamped at the equilibrium geometry $\mathbf{X}_0$. However, the complete Born-Oppenheimer (electric) field-free states are vibronic products of the form:

$$|K, k\rangle = \phi_K(\mathbf{X}; \mathbf{x})\chi_k^K(\mathbf{X}) \qquad (2.32)$$

where $\chi_k^K(\mathbf{X})$ is the solution of the vibrational Schrödinger equation

$$\left[\hat{T}_n(\mathbf{X}) + E_K(\mathbf{X})\right]\chi_k^K(\mathbf{X}) = E_k^K \chi_k^K(\mathbf{X}). \qquad (2.33)$$

Note that all quantities in Eq. (2.33) are field-free even though the superscript $^{(0)}$ has been omitted. $\hat{T}_n$ is the vibrational kinetic energy operator; the electronic energy $E_K = E_K^{(0)}$ [cf. Eq. (2.2)] serves as the potential energy function for vibrational motion in electronic state $|K\rangle$; and $E_k^K = E_K(\mathbf{X}_0) + e_k^K$ is the total (vibronic) energy of state $|Kk\rangle$.

SOS expressions for the NLO properties in terms of the Born-Oppenheimer vibronic states may be derived simply by making the replacements $|K\rangle \rightarrow |K, k\rangle$ and $\omega_K \rightarrow \omega_{Kk} = E_k^K / \hbar$ in Eqs. (2.7), (2.8) and (2.9). Moreover, the prime on the summations should now be understood to omit just the vibronic ground state $|K = 0, k = 0\rangle$. It is important to note that the sums include all $|0, k\rangle$ with k unequal to zero, which correspond to excited vibrational states on the ground electronic state potential energy surface (PES). These are the terms that give rise to the vibrational (hyper)polarizability.

Let us consider the SOS expression for the linear polarizability [cf. Eq. (2.7)] in terms of vibronic states. After integration over electronic coordinates the dipole moment matrix element $\langle 0, 0 | \hat{\mu}_\zeta(\mathbf{x}, \mathbf{X}) | K = 0, k \rangle$ becomes $\langle 0 | \mu_\zeta^{K=0}(\mathbf{X}) | k \rangle$, where $\mu_\zeta^{K=0}(\mathbf{X})$ is the dipole moment *function* (ζ component) for electronic state $|K = 0\rangle$. The vibrational polarizability can, then, be written as

$$\alpha_{\zeta,\eta}^{v}(-\omega_\sigma; \omega_1) = \frac{1}{\hbar} \sum P_{-\sigma,1} \sum_k' \frac{1}{\omega_k - \omega_\sigma} \langle 0 | \mu_\zeta(\mathbf{X}) | k \rangle \langle k | \mu_\eta(\mathbf{X}) | 0 \rangle \qquad (2.34)$$

where, for convenience, the superscript $K = 0$ on μ has been omitted.

At this point it is convenient to introduce vibrational normal coordinates $\{Q_a\}$. This implies a transformation from the $3N$ displacement coordinates $\mathbf{X} - \mathbf{X}_0$ to $3N - 6 Q_a$ plus three center-of-mass coordinates and three angles describing the molecular orientation (for linear molecules there are $3N - 5$ normal coordinates and two angles of orientation). Assuming small displacements about the equilibrium geometry, as well as fixed orientation (see more later) and center of mass, we may expand the dipole moment function in the power series:

$$\mu_\zeta(\mathbf{X}) = \mu_\zeta(\mathbf{X}_0) + \sum_a \left(\frac{\partial \mu_\zeta}{\partial Q_a}\right)_0 Q_a + \frac{1}{2} \sum_{a,b} \left(\frac{\partial^2 \mu_\zeta}{\partial Q_a \partial Q_b}\right)_0 Q_a Q_b + \cdots, \quad (2.35)$$

with the sums running over all normal coordinates. Upon evaluation of the matrix elements in Eq. (2.34) by integration over normal coordinates, the contribution due to the constant term in Eq. (2.35) will vanish because the wavefunctions are orthogonal. Hence, the first term to consider is linear in the displacements. It may be referred to as the harmonic, or zeroth-order, electric dipole contribution. The term that is quadratic in the displacements is defined to be first-order in electric dipole anharmonicity, the cubic term is second-order, and so forth.

A similar expansion of the PES leads to vanishing first derivatives (due to the equilibrium condition at $\mathbf{X} = \mathbf{X}_0$):

$$E_{K=0}(\mathbf{X}) = E_{K=0}(\mathbf{X}_0) + V_n(\mathbf{Q})$$

$$V_n(\mathbf{Q}) = \frac{1}{2} \sum_{a,b} \left(\frac{\partial^2 V_n}{\partial Q_a \partial Q_b}\right)_0 Q_a Q_b$$

$$+ \frac{1}{6} \sum_{a,b,c} \left(\frac{\partial^3 V_n}{\partial Q_a \partial Q_b \partial Q_c}\right)_0 Q_a Q_b Q_c + \cdots \qquad (2.36)$$

Thus, the quadratic terms constitute the mechanical zeroth-order (harmonic) approximation and the second derivatives are the harmonic (or quadratic) force constants. The cubic terms, involving cubic force constants, are first-order in mechanical anharmonicity, etc. The expansions in Eqs. (2.35) and (2.36) form the basis for the double (electrical, mechanical) perturbation theory treatment of vibrational (hyper)polarizabilities developed by Bishop and Kirtman [4, 21], hereafter referred to as BKPT. In BKPT the vibrational wavefunctions, $\chi_k(\mathbf{X})$, and vibrational energy levels, e_k, are found by ordinary Rayleigh-Schrödinger perturbation theory applied to the vibrational Schrödinger equation (2.33) using the potential energy function of Eq. (2.36) with the definition of orders given above (the $E_{K=0}(\mathbf{X}_0)$ term appears on both sides of Eq.(2.33) and cancels out).

Substituting the dipole moment and potential energy expansions into Eq. (2.34), and assuming that the frequency of the electric field is in the non-resonant region of the spectrum (below the lowest electronic transition), the result can be expressed as the perturbation series

$$\alpha^{\mathrm{v}}(-\omega_\sigma; \omega) = \left[\mu^2\right] = \left[\mu^2\right]^0 + \left[\mu^2\right]^{\mathrm{II}} + \cdots \qquad (2.37)$$

For sake of simplicity, the directional indices in Eq. (2.37) have been removed as well as the optical frequency on the right hand side. We have also avoided writing out the full expressions, which may be found in [4, 21]. The superscripts indicate the total order in mechanical and electrical anharmonicity as explained below. Our notation μ^2 indicates that the electrical factor in each term is a product of two dipole derivatives. The product $\frac{\partial\mu}{\partial Q_a}\frac{\partial\mu}{\partial Q_b}$ is zeroth-order in electrical anharmonicity, $\frac{\partial^2\mu}{\partial Q_a\partial Q_b}\frac{\partial\mu}{\partial Q_c}$ is first-order, $\frac{\partial^2\mu}{\partial Q_a\partial Q_b}\frac{\partial^2\mu}{\partial Q_c\partial Q_d}$ (and $\frac{\partial^3\mu}{\partial Q_a\partial Q_b\partial Q_c}\frac{\partial\mu}{\partial Q_d}$) is second-order, and so forth.

Each of the individual perturbation terms within the square brackets of Eq. (2.37) is the product of an electric dipole factor, as just described, multiplied by a mechanical factor that depends on the vibrational force constants. After carrying out the perturbation expansion, the mechanical factors contain an harmonic frequency in the denominator and anharmonic force constants in the numerator. In zeroth-order the mechanical factor in the numerator is unity [see Eq. (2.38) below]; in first-order the individual terms are linear in the cubic force constants; in second-order they are quadratic in the cubic force constants or linear in the quartic force constants, etc.

We have used superscripts on the square brackets to specify the total order of perturbation theory, which is the sum of the order in electrical anharmonicity, $n = 0, 1, \ldots$ plus the order in mechanical anharmonicity, $m = 0, 1, \ldots$ This means that the term of order II contains all contributions of order (n, m) such that $n + m = 2$. When the anharmonicity is small, the zeroth-order doubly harmonic approximation ($\omega_\sigma = \omega_1 = \omega$):

$$\left[\mu^2\right]^0 = \frac{1}{2}\sum_\sigma P_{\sigma,1}\sum_a \frac{\partial\mu_\zeta}{\partial Q_a}\frac{\partial\mu_\eta}{\partial Q_a}\left(\frac{1}{\omega_a - \omega_\sigma}\right)\left(\frac{1}{\omega_a + \omega_\sigma}\right) \qquad (2.38)$$

may be sufficient. Some manipulations, and the fact that the matrix elements of Q_a depend upon ω_a, are required to achieve the form of Eq. (2.38) (see [4]). Finally, we note that a symmetry rule prevents any odd order perturbation terms from occurring in Eq. (2.37).

Next we turn to the first hyperpolarizability. Reflecting on Eq. (2.8) one can see that there are three ways to generate the ground electronic state: (1) $K = 0$, $L \neq 0$; (2) $K \neq 0$, $L = 0$; and (3) $K = 0$, $L = 0$. For case (1) we neglect the vibrational energy associated with excited electronic state L as compared to the electronic excitation energy. Then the sum over L creates the linear polarizability. If the optical frequencies associated with the electric fields lie well below the first electronic transition the frequency-dependence of the linear polarizability may be neglected (see further discussion later). In that event, the analogue of Eq. (2.37) for case 1 is

$$\beta^{\mathrm{v}}(-\omega_\sigma; \omega_1, \omega_2) = [\mu\alpha] = [\mu\alpha]^0 + [\mu\alpha]^{\mathrm{II}} + \cdots \quad \text{case (1).} \qquad (2.39)$$

In order to obtain this result both α and μ have been expanded as power series in the normal coordinates. The square bracket $[\mu\alpha]$ indicates that each term involves the product of a dipole derivative multiplied by a linear polarizability derivative. Using the same definition of orders as for the dipole expansion it turns out, again, that only even order terms appear in the perturbation series.

For case (2) identical considerations apply as for case (1). In fact, the contribution to the vibrational first hyperpolarizability is the same for both. That leaves $K = L = 0$, which gives rise to:

$$\beta^{\mathrm{v}}(-\omega_\sigma; \omega_1, \omega_2) = \left[\mu^3\right] = \left[\mu^3\right]^{\mathrm{I}} + \left[\mu^3\right]^{\mathrm{III}} + \cdots \quad \text{case (3).} \qquad (2.40)$$

At this point it should be obvious how to interpret the square brackets in Eq. (2.40).

Finally, we consider the second (hyper)polarizability [cf. Eq. (2.9)]. There are four different types of square bracket that occur. Their form and origin are shown below:

$$
\begin{aligned}
[\mu\beta]^0 + [\mu\beta]^{\mathrm{II}} + \cdots \quad &\longleftarrow \quad \gamma^{(+)}(K = 0; M = 0) \\
\left[\alpha^2\right]^0 + \left[\alpha^2\right]^{\mathrm{II}} + \cdots \quad &\longleftarrow \quad \gamma^{(+)}(L = 0) \\
\left[\mu^2\alpha\right]^{\mathrm{I}} + \left[\mu^2\alpha\right]^{\mathrm{III}} + \cdots \quad &\longleftarrow \quad \begin{cases} \gamma^{(+)}(K, L = 0; K, M = 0; L, M = 0) \\ \gamma^{(-)}(K = 0; L = 0) \end{cases} \\
\left[\mu^4\right]^{\mathrm{II}} + \left[\mu^4\right]^{\mathrm{IV}} + \cdots \quad &\longleftarrow \quad \gamma^{(+)}(K, L, M = 0); \gamma^{(-)}(K = 0; L = 0).
\end{aligned}
$$

$$(2.41)$$

Here $\gamma^{(+)}$ and $\gamma^{(-)}$ refer to the first and second terms in Eq. (2.9) respectively. The semi-colons inside the parentheses separate the different cases, which are identified by specifying the electronic indices that are zero (while the others take all values except zero). The $\left[\mu^4\right]$ perturbation series begins at second-order because the zeroth-order terms from $\gamma^{(+)}$ and $\gamma^{(-)}$ cancel one another [22].

From Eqs. (2.35) and (2.36) it can be inferred that ab initio computations will rapidly become more time consuming as one proceeds to higher order in perturbation theory because of the occurrence of higher order derivatives, which are more difficult to obtain individually and more numerous. Even a double harmonic treatment can be tedious for a large molecule since all the harmonic force constants must be calculated. Thus, it is of value to have an alternative procedure that is computationally more efficient, even though some (reasonable) approximations may have to be introduced. The so-called finite field-nuclear relaxation (FF-NR) method fulfills that goal ([23], see also [24]). It has been successfully applied to many small-to-medium size molecules and, recently, to infinite periodic systems as well [25, 26]. Although there is a subsequent more advanced version (see later), the original method is, with one exception, equivalent to BKPT through first-order. In either procedure one obtains the leading term in the perturbation series for each type of square bracket. For the static γ^v, this means that a second-order $\left[\mu^4\right]^{II}$ term in Eq. (2.41) is also included. A limitation of the FF-NR method is that the L&NLO properties are determined only in those circumstances where all external fields are either static or near the high frequency limit. This covers several of the more important cases. A modification is required to treat DFWM [27].

The main step in the FF-NR procedure is geometry optimization in the presence of a finite (static) field. In the simpler (first-order) version this is followed by evaluation of the static electronic dipolar properties (μ^e, α^e, β^e) at the relaxed geometry. For good accuracy the geometry optimization should be done with tight thresholds. Most importantly, the molecule cannot be allowed to rotate (in this regard the field-free Eckart conditions must be satisfied [28]) so that the direction of the field with respect to molecular axes remains unchanged during geometry optimization. The effect of rotation may be taken into account by carrying out calculations for different field directions followed by classical orientational averaging [29].

Let P^e be a static electronic dipolar property and Q_E the optimized set of normal coordinate displacements in the presence of the finite field **E**. Then, if $\Delta P^e = P^e(\mathbf{E}, Q_E) - P^e(\mathbf{E}, 0)$ is expanded as a power series in **E**, the expansion coefficients can be simply related to the vibrational (hyper)polarizabilities at $Q = 0$, i.e.

$$(\Delta\mu_\zeta)_{Q_E} = \sum_\eta a_1 E_\eta + \frac{1}{2}\sum_{\eta\kappa} b_1 E_\eta E_\kappa + \frac{1}{6}\sum_{\eta\kappa\lambda} g_1 E_\eta E_\kappa E_\lambda + \cdots$$

$$(\Delta\alpha_{\zeta\eta})_{Q_E} = \sum_\kappa b_2 E_\kappa + \frac{1}{2}\sum_{\kappa\lambda} g_2 E_\kappa E_\lambda + \cdots$$

$$(\Delta\beta_{\zeta\eta\kappa})_{Q_E} = \sum_\lambda g_3 E_\lambda \tag{2.42}$$

with

$$a_1 - \alpha^e(0) = \alpha_{\zeta\eta}^{nr}(0;0); \quad b_1 - \beta^e(0) = \beta_{\zeta\eta\kappa}^{nr}(0;0,0); \quad g_1 - \gamma^e(0) = \gamma_{\zeta\eta\kappa\lambda}^{nr}(0;0,0,0)$$
$$b_2 - \beta^e(0) = \beta_{\zeta\eta\kappa}^{nr}(-\omega;\omega,0)|_{\omega\to\infty}; \quad g_2 - \gamma^e(0) = \gamma_{\zeta\eta\kappa\lambda}^{nr}(-\omega;\omega,0,0)|_{\omega\to\infty}$$
$$g_3 - \gamma^e(0) = \gamma_{\zeta\eta\kappa\lambda}^{nr}(-2\omega;\omega,\omega,0)|_{\omega\to\infty}$$

$$(2.43)$$

and $\alpha^e(0)$, $\beta^e(0)$, $\gamma^e(0)$ equal to the static electronic α, β, γ. The superscript nr here refers to the first-order nuclear relaxation treatment. Although all calculations are done with static fields, the last two lines in Eq. (2.43) yield dynamic NLO properties—the subscript $\omega \to \infty$ implies the limiting high frequency value. This same value is obtained from BKPT if the quantity $\left(\frac{\omega_v}{\omega}\right)^2$, with ω_v equal to a fundamental vibrational frequency, is negligible compared to unity for all ω_v. Under the same approximation the nr contribution to second and third harmonic generation, namely $\beta_{\zeta\eta\kappa}^{nr}(-2\omega;\omega,\omega)$ and $\gamma_{\zeta\eta\kappa\lambda}^{nr}(-3\omega;\omega,\omega,\omega)$, is zero. In fact, as a general rule, the more static fields that define the process, the larger will be the nuclear relaxation contribution relative to the pure electronic term. This means, for example, that one would expect vibrations to be more important for $\gamma(-\omega;\omega,0,0)$ (dc-Kerr effect) than for $\gamma(-2\omega;\omega,\omega,0)$ or $\beta(-\omega;\omega,0)$. In fact, for dc-Kerr the vibrations may be more important than pure electronic motions and for static γ vibrational contributions are often dominant.

The intensity-dependent refractive index (IDRI) $\gamma(-\omega;\omega,\omega,-\omega)$, or degenerate four-wave mixing (DFWM), is a special case. It turns out that the vibrational contribution is quite important for this property because one of the frequencies occurs with a negative sign leading to a cancellation that effectively produces two static fields. As noted above the original FF-NR method is readily modified to calculate DFWM [27].

Closely related to the FF-NR procedure are methods based on what are known as field-induced coordinates (FICs) [30, 31]. These coordinates are determined by the optimized set of normal coordinate displacements Q_E, defined above. After expanding the latter as a power series in the field, the linear coefficients (see below) determine three first-order FICs, one for each Cartesian field direction. The quadratic coefficients yield the second-order FICs, of which there are six—one for each pair of field directions. Regardless of the size of the molecule these are the only (9) coordinates that are needed to obtain the first-order vibrational NLO properties. They can be found analytically as well, but the numerical procedure is more efficient for large molecules.

Next we write the first-order FIC (κ component) as

$$\bar{\chi}_1^\kappa = \sum_a \left(M_1^\kappa\right)_a Q_a, \tag{2.44}$$

where $\left(M_1^\kappa\right)_a$ is the linear coefficient in the expansion of $(Q_a)_E$, as a function of E, obtained by a numerical fit. Then, in terms of the $\left(M_1^\kappa\right)_a$ the nr EOPE, for example, becomes just

$$\beta^{nr}_{\zeta\eta\kappa}(-\omega;\omega,0)_{\omega\to\infty} = \frac{1}{2}\frac{\partial\alpha_{\zeta\eta}}{\partial\bar{\chi}^{\kappa}_{1}}\sum_{a}\left(M^{\kappa}_{1}\right)^{2}_{a}. \tag{2.45}$$

In Eq. (2.45) the derivative in front of the summation must be evaluated numerically as well. However, only a single derivative is involved rather than a separate derivative for each normal coordinate. Moreover, symmetry coordinates may be used everywhere, instead of normal coordinates, making it unnecessary to calculate the Hessian.

An expression similar to Eq. (2.45) can be written for $\gamma^{nr}(-2\omega;\omega,\omega,0)_{\omega\to\infty}$ except that α is replaced by β on the right hand side and the factor of $1/2$ is replaced by $1/6$. In this case, as well as in Eq. (2.45), only the first-order FIC is needed because the anharmonic contributions to EFISHG vanish in the limit $\omega\to\infty$. The expressions for the remaining nr properties are somewhat more complicated, but similar simplifications occur. The static γ is the most difficult property to compute since both first- and second-order FICs contribute. Moreover, several anharmonicity parameters enter into the formulas. A complete set of expressions for static and dynamic vibrational properties is given in [30].

The FF-NR approach has been extended beyond the lowest-order square bracket terms of each type. In fact, a treatment that is exact in principle is available [24]. One simply replaces the electronic property values in Eq. (2.42) by their zero-point vibrational average (ZPVA). In practice, the accuracy obtained will depend upon the level of approximation used to compute the ZPVA. Although this procedure has been successfully applied to small molecules, and is currently moving forward, further developments are necessary before it can be applied to large systems. It is important to realize, however, that the resulting contributions can be quite significant in systems with low frequency, large amplitude vibrational modes. The possibility of treating just that limited subset of vibrations within a large system has begun to be explored [32].

In the FF-NR and BKPT methods the electronic transition frequencies are assumed to be much larger than the frequencies of the (external) laser optical fields. Hansen et al. [33] have developed a response theory formulation that accounts for the vibrational contribution which is thereby omitted. This so-called 'mixed' term is difficult to compute, but could sometimes be important. Nonetheless, it is zero both in the static limit and in the 'infinite' optical frequency limit previously defined [34]. Initial calculations carried out by Hansen et al. [33] found the mixed term to be small.

References

1. Champagne, B.: Polarizabilities and hyperpolarizabilities. In: Springborg, M. (ed.) Chemical Modelling: Applications and Theory, pp. 43–88. Royal Society of Chemistry, Cambridge (2010)
2. Bishop, D.M., Kirtman, B., Champagne, B.: Differences between the exact sum-over-states and the canonical approximation for the calculation of static and dynamic hyperpolarizabilities. J. Chem. Phys. **107**, 5780–5787 (1997)
3. Marks, T.J., Ratner, M.A.: Design, synthesis, and properties of molecule-based assemblies with large second-order optical nonlinearities. Ang. Chem. Int. Ed. **34**, 155–173 (1995)

4. Bishop, D.M., Kirtman, B.: A perturbation method for calculating vibrational dynamic dipole polarizabilities and hyperpolarizabilities. J. Chem. Phys. **95**, 2646–2658 (1991)
5. Orr, B.J., Ward, J.F.: Perturbation theory of the non-linear optical polarization of an isolated system. Mol. Phys. **20**, 513–526 (1971)
6. Kuzyk, M.G.: Physical limits on electronic nonlinear molecular susceptibilities. Phys. Rev. Lett. **90**, 039902 (2003). (Phys. Rev. Lett. **85**, 001218, (2000). Erratum)
7. Champagne, B., Kirtman, B.: Comment on 'Physical limits on electronic nonlinear molecular susceptibilities'. Phys. Rev. Lett. **95**, 109401 (2005)
8. Sekino, H., Bartlett, R.J.: Frequency dependent nonlinear optical properties of molecules. J. Chem. Phys. **85**, 976–989 (1986)
9. Karna, S.P., Dupuis, M.: Frequency dependent nonlinear optical properties of molecules: formulation and implementation in the HONDO program. J. Comp. Chem. **12**, 487–504 (1991)
10. Linderberg, J., Ohrn, Y.: Propagators in quantum chemistry. Academic Press, New York (1973)
11. Nielsen, E.S., Jørgensen, P., Oddershede, J.: Transition moments and dynamic polarizabilities in a second order polarization propagator approach. J. Chem. Phys. **73**, 6238–6246 (1980)
12. Olsen, J., Jørgensen, P., Helgaker, T., Oddershede, J.: Quadratic response functions in a second-order polarization propagator framework. J. Phys. Chem. A **109**, 11618–11628 (2005)
13. Bishop, D.M., De Kee, D.W.: The frequency dependence of nonlinear optical processes. J. Chem. Phys. **104**, 9876–9887 (1996)
14. Runge, E., Gross, E.K.U.: Density-functional theory for time-dependent systems. Phys. Rev. Lett. **52**, 997–1000 (1984)
15. Pople, J.A., Gill, P.M.W., Johnson, B.G.: Kohn-Sham density-functional theory within a finite basis set. Chem. Phys. Lett. **199**, 557–560 (1992)
16. Orlando, R., Lacivita, V., Bast, R., Ruud, K.: Calculation of the first static hyperpolarizability tensor of three-dimensional periodic compounds with a local basis set: a comparison of LDA, PBE, PBE0, B3LYP, and HF results. J. Chem. Phys. **132**, 244106 (2010)
17. Orlando, R., Bast, R., Ruud, K., Ekström, E., Ferrabone, M., Kirtman, B., Dovesi, R.: The first and second static electronic hyperpolarizabilities of zigzag boron nitride nanotubes. An ab initio approach through the coupled perturbed Kohn-Sham scheme. J. Phys. Chem. A **115**, 12631–12637 (2011)
18. Ekström, U., Visscher, L., Bast, R., Thorvaldsen, A.J., Ruud, K.: Arbitrary-order density functional response theory from automatic differentiation. J. Chem. Theory Comput. **6**, 1971–1980 (2010)
19. Ekström, U.: XCFun library. http://www.admol.org/xcfun (2010)
20. Shedge, S.V., Carmona-Espíndola, J., Pal, S., Köster, A.M.: Comparison of the auxiliary density perturbation theory and the noniterative approximation to the coupled perturbed Kohn-sham method: case study of the polarizabilities of disubstituted azoarene molecules. J. Phys. Chem. A **114**, 2357–2364 (2010)
21. Bishop, D.M., Luis, J.M., Kirtman, B.: Additional compact formulas for vibrational dynamic dipole polarizabilities and hyperpolarizabilities. J. Chem. Phys. **108**, 10013–10017 (1998)
22. Kirtman, B., Bishop, D.M.: Evaluation of vibrational hyperpolarizabilities. Chem. Phys. Lett. **175**, 601–607 (1990)
23. Bishop, D.M., Hasan, M., Kirtman, B.: A simple method for determining approximate static and dynamic vibrational hyperpolarizabilities. J. Chem. Phys. **103**, 4157–4159 (1995)
24. Kirtman, B., Luis, J.M., Bishop, D.M.: Simple finite field method for calculation of static and dynamic vibrational hyperpolarizabilities: curvature contributions. J. Chem. Phys. **108**, 10008–10012 (1998)
25. Ferrabone, M., Kirtman, B., Lacivita, V., Rérat, M., Orlando, R., Dovesi, R.: Vibrational contribution to static and dynamic (hyper)polarizabilities of zigzag BN nanotubes calculated by the finite field nuclear relaxation method. Int. J. Quant. Chem. **112**, 2160–2170 (2012)
26. Lacivita, V., Rérat, M., Kirtman, B., Orlando, R., Ferrabone, M., Dovesi, R.: Static and dynamic coupled perturbed Hartree-Fock vibrational (hyper)polarizabilities of polyacetylene calculated by the finite field nuclear relaxation method. J. Chem. Phys. **137**, 014103 (2012)

27. Kirtman, B., Luis, J.M.: Simple finite field nuclear relaxation method for calculating vibrational contribution to degenerate four-wave mixing. J. Chem. Phys. **128**, 114101 (2008)
28. Luis, J.M., Duran, M., Andrés, J.L., Champagne, B., Kirtman, B.: Finite field treatment of vibrational polarizabilities and hyperpolarizabilities: on the role of the Eckart conditions, their implementation, and their use in characterizing key vibrations. J. Chem. Phys. **111**, 875–884 (1999)
29. Bishop, D.M., Lam, B., Epstein, S.T.: The Stark effect and polarizabilities for a diatomic molecule. J. Chem. Phys. **88**, 337–341 (1988)
30. Luis, J.M., Duran, M., Champagne, B., Kirtman, B.: Determination of vibrational polarizabilities and hyperpolarizabilities using field-induced coordinates. J. Chem. Phys. **113**, 5203–5213 (2000)
31. Luis, J.M., Duran, M., Kirtman, B.: Field-induced coordinates for the determination of dynamic vibrational nonlinear optical properties. J. Chem. Phys. **115**, 4473–4483 (2001)
32. Garcia-Borràs, M., Solà, M., Lauvergnat, D., Reis, H., Luis, J.M., Kirtman, B.: A full dimensionality approach to evaluate the nonlinear optical properties of molecules with large amplitude anharmonic tunneling motions. J. Chem. Theory Comput. **9**, 520–532 (2013). (See also earlier papers cited therein)
33. Hansen, M.B., Christiansen, O., Hättig, C.: Automated calculation of anharmonic vibrational contributions to first hyperpolarizabilities: quadratic response functions from vibrational configuration interaction wave functions. J. Chem. Phys. **131**, 154101 (2009)
34. Kirtman, B., Luis, J.M.: On the contribution of mixed terms in response function treatment of vibrational nonlinear optical properties. Int. J. Quant. Chem. **111**, 839–847 (2011)

Chapter 3
Quantum-Mechanical Treatment of Responses to Electric Fields—Extended Systems

Abstract In this chapter we discuss the quantum-mechanical treatment of extended systems exposed to electric fields. Special emphasis is put on systems that are regular and so large that they can be considered as being infinite and periodic. Moreover, it is demonstrated that even the linear responses to static electric fields depend on the surfaces of the systems, although these need not be included directly in the calculations. Also perturbation-theoretical treatments of linear and non-linear responses to oscillatory fields are discussed. We consider both the vector-potential and the scalar-potential treatment of the electric fields.

3.1 Introduction

In the preceding chapter we discussed quantum-mechanical methods for calculating the L&NLO properties of ordinary molecules with an eye towards their extension to macromolecular systems. Now, in this chapter, we want to approach the treatment of macromolecular systems from the opposite point of view, namely from the perspective of idealized infinite periodic systems. The infinite periodic model is relevant for systems that consist of a large number of regularly placed, identical units (with deviations from regularity only at the surfaces). The study of such systems allows us to raise issues that are pertinent for the more general case treated by the elongation method. The latter method is also valid, of course, for periodic systems, although it is especially advantageous when non-periodicity is important. Many interesting physico-chemical problems that involve defects (in a general sense), as well as random or quasi-random order, fall under that rubric.

In Chap. 2 the electric field(s) in $\hat{H}'(\mathbf{X}, \mathbf{x}, t)$ were expressed in terms of the scalar potential. For infinite periodic systems, however, frequency-dependent L&NLO properties are more conveniently treated using the vector potential [1, 2]. Thus, we begin in Sect. 3.2 with a brief discussion of the scalar and vector potentials as related to the choice of the gauge. Then, in Sect. 3.3 we turn to the theoretical treatment of the electronic response to electrostatic fields by regular systems large enough to have reached the thermodynamic limit. In that limit the response properties are either proportional to, or independent of, the system size. In Sect. 3.4 we demonstrate how

© The Author(s) 2015
F.L. Gu et al., *Calculations on nonlinear optical properties for large systems*,
SpringerBriefs in Electrical and Magnetic Properties of Atoms,
Molecules, and Clusters, DOI 10.1007/978-3-319-11068-4_3

the infinite, periodic system exposed to an electrostatic field can be treated using either the vector or the scalar potential. Finally, the changes required to convert the ab initio TDHF and TDDFT perturbation equations for a finite system, based on the scalar potential, to those for an infinite periodic system, based on the vector potential, are described in Sect. 3.5.

3.2 The Choice of the Gauge: Scalar and Vector Potentials

For an electron in an external electromagnetic field, the time-dependent KS-DFT single-particle equation takes the form

$$
\left\{ \frac{1}{2m} \left[-i\hbar\nabla + \frac{e}{c}\mathbf{A}_{\text{ext,E}}(\mathbf{r}, t) \right]^2 + V_{\text{ext,n}}(\mathbf{r}) + V_{\text{J}}(\mathbf{r}) + V_{\text{xc}}(\mathbf{r}, t) \right.
$$

$$
\left. + V_{\text{ext,E}}(\mathbf{r}, t) \right\} \psi_i(\mathbf{x}, t) = i\hbar \frac{\partial}{\partial t} \psi_i(\mathbf{x}, t). \tag{3.1}
$$

Here, m is the mass of the electron and c is the speed of light. V_{J} is the coulombic potential due to electron-electron repulsion including self-interaction; exchange and correlation effects are contained in V_{xc}, which is a functional of the (time-dependent) electron density; and $V_{\text{ext;n}}$ accounts for the electrostatic nuclear-electron attraction. The effects of the electric field are described via the vector potential $\mathbf{A}_{\text{ext,E}}$ and the scalar potential $V_{\text{ext;E}}$. Exactly the same terms appear in the time-dependent Hartree-Fock approach except that V_{xc} becomes an orbital-dependent operator and correlation effects are not included [see Eq. (2.14) for the case $\mathbf{A}_{\text{ext;E}} = \mathbf{0}$, $V_{\text{ext;E}} = 0$]. The electric field, $\mathbf{E}$, and the magnetic field, $\mathbf{B}$, are given through

$$
\mathbf{E}(\mathbf{r}, t) = -\nabla V_{\text{ext,E}}(\mathbf{r}, t) - \frac{1}{c}\frac{\partial}{\partial t}\mathbf{A}_{\text{ext,E}}(\mathbf{r}, t)
$$

$$
\mathbf{B}(\mathbf{r}, t) = \nabla \times \mathbf{A}_{\text{ext,E}}(\mathbf{r}, t). \tag{3.2}
$$

In the present discussion we shall not be concerned with magnetic fields, so that $\mathbf{B} = \mathbf{0}$, which can be satisfied by taking

$$
\mathbf{A}_{\text{ext,E}}(\mathbf{r}, t) = \nabla f_{\text{ext}}(\mathbf{r}, t), \tag{3.3}
$$

where $f_{\text{ext}}(\mathbf{r}, t)$ is some function. Even then, however, the separation of $\mathbf{E}$ into scalar and vector potential contributions is not unique: different choices define different gauges.

An example that we shall discuss below in some detail is that of a homogeneous, electrostatic field given by a constant $\mathbf{E}$ everywhere in space. This can be described, for example, through a pure scalar potential,

$$V_{ext,E}(\mathbf{r}, t) = -\mathbf{E} \cdot \mathbf{r}$$
$$\mathbf{A}_{ext,E}(\mathbf{r}, t) = \mathbf{0}, \tag{3.4}$$

or, equivalently, through a pure vector potential,

$$V_{ext,E}(\mathbf{r}, t) = 0$$
$$\mathbf{A}_{ext,E}(\mathbf{r}, t) = -c\mathbf{E} \cdot t. \tag{3.5}$$

Linear combinations of the two cases are possible, too, but here we shall only be concerned with those two choices. Equation (3.1) depends on which of the two forms in Eqs. (3.4) and (3.5) is used but, of course, physical observables must be independent of that choice. Below we shall exploit both possibilities.

3.3 Thermodynamic Limit

In this section we begin by considering the theoretical treatment for the response of large, regular systems to electrostatic fields. Our discussion will be based on a scalar potential description of the electrostatic field. In particular, we shall emphasize systems that are so large the thermodynamic limit has been reached for the L&NLO properties of interest. Subsequently, the changes that occur when treating the system as being infinite and periodic will be discussed. It may or may not be feasible to accurately describe such systems by the elongation method without requiring extrapolation. Nonetheless, their analysis gives interesting additional information. In the periodic case, we discuss both the scalar and vector potentials and, moreover, demonstrate how the latter leads to a natural approach for the treatment of time-dependent fields.

A schematic example of systems like those we shall consider in this section is given in Fig. 3.1. The figure shows a quasi-two-dimensional system with a rectangular shape that is assumed to be neutral. The central part, i.e. the so-called central region, consists of a large number of identical units, each having the size of the crystallographic unit cell. Near the boundaries, deviations from this regularity may show up whereby the equivalent units become larger, as indicated in the figure. Nevertheless, along each of the four sides (in our case), the (larger) units are equivalent, Only at the four corners (again, in our case) may deviations from this regularity be found. The reasons for the larger unit cells along the surfaces can be surface reconstructions or passivating ligands.

If we use the scalar potential to describe the interaction with the electrostatic field, then the total energy due to the interaction with both nuclei and electrons can be written as

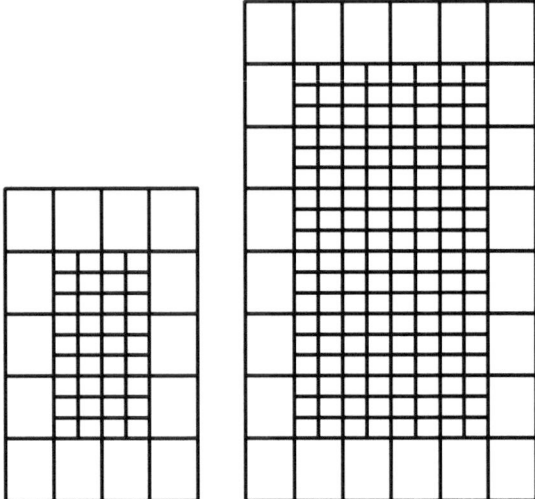

Fig. 3.1 Schematic presentation of a quasi-2D system. The system is separated into a central region, side regions, and corner regions. The changes upon moving from the *left panel* to the *right* one demonstrate what occurs when the system gradually is made larger. For further details, see the text

$$- \mathbf{E} \cdot \left[\sum_m Z_m e \mathbf{R}_m - e \sum_j \langle \psi_j | \mathbf{r} | \psi_j \rangle \right] = \mathbf{E} \cdot \int \mathbf{r} \rho_{\text{tot}}(\mathbf{r}) d\mathbf{r} = -\mathbf{E} \cdot \mu_{\text{tot}}, \quad (3.6)$$

where $Z_m e$ and $\mathbf{R}_m$ are the charge and position, respectively, of the mth nucleus, while $\psi_j(\mathbf{r})$ is the jth electronic orbital. The quantity

$$\rho_{\text{tot}}(\mathbf{r}) = \sum_m e Z_m \delta(\mathbf{r} - \mathbf{R}_m) - e \sum_j |\psi_j(\mathbf{r})|^2 \quad (3.7)$$

is the total charge density from both nuclei and electrons and, finally, μ_{tot} is the total dipole moment. For the present discussion, the dipole moment is the key quantity.

For macroscopic systems sufficiently large that the thermodynamic limit has been reached, the units of a certain region (say, the top side region) in the left hand part of Fig. 3.1 will be identical to those of the same region in the right hand part of the figure. Thus, each of these units will have the same charge independent of the system size. By applying the same argument it follows that the units of the central region are neutral in order to preserve overall neutrality. For the same reason, the sum of the surface charges will vanish.

Imagine now that the shape of the system of Fig. 3.1 is changed. For instance, it may have more sides, or the sides may no longer be pairwise parallel. Even if the system with the new shape and that of Fig. 3.1 have one side in common, the units of this side may have different charges in the two situations. In fact, that must be the case, in general, for the charge neutrality condition to be satisfied. These circumstances are independent of the size of the system as long as it is sufficiently large. In total, we end up with the situation that the charge on each surface unit of a macroscopic sample depends on all the other surface units.

For a macroscopic sample, the surface regions make up a very small part of the total volume and one may, accordingly, suggest that the effect of these regions is negligible for any intensive property of the sample. This is, indeed, very often the case but it is not so for the dipole moment per unit as we now demonstrate. To that end, for very large systems it is convenient to define an intensive property, $\bar{\xi}$, corresponding to each extensive property, ξ,

$$\bar{\xi} = \lim_{N \to \infty} \frac{\xi(N)}{N} = \lim_{N \to \infty} \frac{1}{\Delta N} \left[\xi(N + \Delta N) - \xi(N) \right]. \tag{3.8}$$

In Eq. (3.8) N quantifies the size of the system which, in our case, will be the number of units (here, the unit size is taken to be that of a central unit, even if units at the boundaries may be larger). In the current context ξ is the dipole moment of the system and, accordingly, the dipole moment per unit is defined as the difference in dipole moment per additional unit cell when going from the left part to the right part of Fig. 3.1.

In order to evaluate $\bar{\mu}$ it is convenient to split the integral in Eq. (3.6) into a sum of nine terms from each of the nine regions (i.e., the central region, four side regions and four corners). In doing so we arbitrarily choose the origin of coordinates to be at the middle of the sample. When going from the left part to the right part of Fig. 3.1 the size of the sample is increased by order λ^2 where $\lambda > 1$ is the scaling factor. The number of units in the central region also increases by order λ^2. Since these units are neutral, their contribution to the change in the dipole moment per added unit is the dipole moment of any single unit in that region. The size of each of the four side regions increases by a factor of λ, but the distance of these regions from the origin also increases by a factor of λ. Hence, each side region gives a contribution to the change in the dipole moment per added unit that equals the charge per side unit [see text below Eq. (3.7)] divided by the size of such a unit (measured in number of central unit cells) times a vector that is an outwards pointing normal to the side having the length of the unit cell. Finally, since the corner regions do not change size, they do not contribute to the dipole moment per unit.

In total we see that the dipole moment per unit for a macroscopic sample consists of two contributions. One comes from the dipole moment of a single central unit cell, and one is due to charges at the surfaces. Since the dipole moment (per unit) is the quantity that determines the measured responses to an electric field, it is obvious that the surfaces contribute.

Quasi-one-dimensional systems are extended in one direction and finite in the other two. They have additional interesting features that deserve to be noted. For sufficiently large systems, the dipole moment per unit is the sum of a central unit cell contribution plus the charge at one of the two terminations times the lattice constant. One might, at first, assume that this second term can change essentially arbitrarily through chemical modifications at the terminations. However, it has been shown [3–5] that the charge at the terminations can change only by an integer times the electronic charge. Thus, long chains differing only in the terminations will have dipole moments that differ, at most, by an integer times the electronic charge times the lattice constant.

For systems of higher dimension, on the other hand, the same arguments as above lead to the conclusion that different samples of the same material can have dipole moments per unit that differ by essentially any value. In all cases, however, there is a non-negligible contribution from the surfaces that depends upon the sample shape and dimensionality.

3.4 Infinite Periodic Systems in Static and Dynamic Electric Fields

For a large, regular system not exposed to an electromagnetic field, it can be computationally advantageous to treat this system as if it were infinite and periodic. Then, the spatial electronic orbitals can be written as Bloch functions and accordingly classified through the wavevector $\mathbf{k}$,

$$\psi_j(\mathbf{k}, \mathbf{r}) = e^{i\mathbf{k}\cdot\mathbf{r}/\hbar} u_j(\mathbf{k}, \mathbf{r}), \tag{3.9}$$

where the index j distinguishes between different orbitals belonging to the same $\mathbf{k}$ and u_j is a lattice-periodic function. The wavevector $\mathbf{k}$ can take any value but for most purposes (including the total energy and the electron density) only the information for $\mathbf{k}$ lying in the so-called first Brillouin zone is relevant. In a standard calculation for an infinite periodic system, the continuous variable $\mathbf{k}$ is replaced by a discrete set of values with equidistant spacing along each direction in reciprocal space. This defines the units of the Born von Kármán (BvK) zone.

When the infinite periodic system is exposed to an electrostatic field, several conceptual problems arise.

First of all, as shown above, the dipole moment, which quantifies the responses to the field, has a contribution from the surfaces. On the other hand, by construction, the infinite periodic system has no surfaces. Second, in either treatment (finite or infinite periodic), the field can induce electron tunneling from one side of the system to the other. Third, the scalar potential breaks translational symmetry, which is at the root of the periodic treatment and, in addition, it is unbounded for an infinite system.

As far as the dipole moment per unit is concerned, for an infinite periodic system this property must be calculated using expressions that differ significantly from those of Eqs. (3.6) and (3.8). Following earlier work by Blount [6], two different time-independent formulations were suggested about 20 years ago by Resta [7] and by King-Smith and Vanderbilt [8]. It was not until about 10 years later that a formal mathematical derivation was given [9].

We, first, present the expression due to Resta. Let $\mathbf{a}$, $\mathbf{b}$, and $\mathbf{c}$ be the primitive lattice vectors in real space and $\mathbf{a}^* = 2\pi \frac{\mathbf{b} \times \mathbf{c}}{\mathbf{a} \cdot (\mathbf{b} \times \mathbf{c})}$, $\mathbf{b}^* = 2\pi \frac{\mathbf{c} \times \mathbf{a}}{\mathbf{b} \cdot (\mathbf{c} \times \mathbf{a})}$, and $\mathbf{c}^* = 2\pi \frac{\mathbf{a} \times \mathbf{b}}{\mathbf{c} \cdot (\mathbf{a} \times \mathbf{b})}$ be the corresponding primitive lattice vectors in reciprocal space. According to Resta [7], the electronic contribution to the dipole moment per unit can, then, be written as

$$\bar{\mu}_e = \bar{\mu}_{e,\mathbf{a}} + \bar{\mu}_{e,\mathbf{b}} + \bar{\mu}_{e,\mathbf{c}} \tag{3.10}$$

in which

$$\bar{\mu}_{e,s} = \mathbf{s}\frac{eK_s}{2\pi K}\text{Im ln det }\underline{\underline{S}}_s^+ = -\mathbf{s}\frac{eK_s}{2\pi K}\text{Im ln det }\underline{\underline{S}}_s^- \equiv \frac{1}{K}\langle\Psi_e|\hat{\mu}_{e,s}|\Psi_e\rangle. \tag{3.11}$$

Here, $\mathbf{s}$ equals $\mathbf{a}$, $\mathbf{b}$, or $\mathbf{c}$; K is the total number of discrete $\mathbf{k}$ points; and K_s is the number along the $\mathbf{s}^*$ direction. The square matrices $\underline{\underline{S}}_s^{\pm}$, which have the dimension of the number of electrons (per neutral unit cell) times K, contain the matrix elements $\langle\psi_{i_1}(\mathbf{k}_1, \mathbf{r})|e^{\pm i\Delta\mathbf{k}_s \cdot \mathbf{r}/\hbar}|\psi_{i_2}(\mathbf{k}_2, \mathbf{r})\rangle$. These matrix elements are non-zero only for $\mathbf{k}_1 - \mathbf{k}_2 = \pm\Delta\mathbf{k}_s$ where $\Delta\mathbf{k}_s$ is the $\mathbf{k}$ spacing in the $\mathbf{s}^*$ direction (and for orbitals of the same spin). In the last formula on the rhs of Eq. (3.11) we have written the expression in terms of the expectation value of an operator (whose precise form is not important here) evaluated using the Slater determinant wavefunction, Ψ_e, for the occupied orbitals of the K units that make up the BvK zone.

There are several interesting points regarding Eq. (3.11). One of them is that det $\underline{\underline{S}}_s^{\pm}$ is non-vanishing only for systems with an energy gap between occupied and unoccupied orbitals. This reflects the fact that, for a metallic system, the electronic polarization is undefined and the polarizability diverges. Another concerns translational symmetry. Whereas the electronic dipole moment operator, $(-e)\mathbf{r}$, which appears in the scalar potential, does not possess the symmetry of the lattice, the operator in Eq. (3.11) does possess the symmetry of the BvK zone. That is to say, it has the periodicity of K unit cells. Finally, Im ln det $\underline{\underline{S}}_s^{\pm}$ is simply the phase of the complex number det $\underline{\underline{S}}_s^{\pm}$. As such it contains an undefined additive integer multiple of 2π. Ultimately, this means that $\bar{\mu}_e$ contains undefined additive contributions that can, actually, be related to the surface contributions mentioned in the previous section. These contributions depend upon the dimensionality, as well as the shape of the surfaces and the direction of the field. For special cases, such as quasi-one-dimensional systems, the additive contributions become an integer multiple of the lattice constant times the elementary charge. Otherwise, they may take any value.

An alternative expression for the electronic contribution to the dipole moment per unit was suggested by King-Smith and Vanderbilt [8], i.e.

$$\bar{\mu}_e = -\frac{ie\hbar}{K}\sum_{\mathbf{k}}\sum_{j}\langle u_j(\mathbf{k}, \mathbf{x})|\nabla_k u_j(\mathbf{k}, \mathbf{r})\rangle$$

$$= -\frac{ie\hbar}{K}\sum_{\mathbf{k}}\sum_{j}\langle e^{-i\mathbf{k}\cdot\mathbf{r}/\hbar}\psi_j(\mathbf{k}, \mathbf{r})|\nabla_k e^{-i\mathbf{k}\cdot\mathbf{r}/\hbar}\psi_j(\mathbf{k}, \mathbf{r})\rangle \equiv \frac{1}{K}\langle\Psi_e|\hat{\mu}_e|\Psi_e\rangle,$$

$$\tag{3.12}$$

where ∇_k is the gradient operator with respect to $\mathbf{k}$. In this case, the operator in the last term of Eq. (3.12) possesses the lattice periodicity. Moreover, Eq. (3.12) can be obtained from Eqs. (3.10) and (3.11) in the limit of a very dense set of $\mathbf{k}$ points. This formulation also leads to undefined additive contributions that, now, can be related to the phases of the crystal orbitals. The interpretation of these contributions is the same as above.

We have noted that the expressions for the electronic dipole moment per unit in Eqs. (3.11) and (3.12) both contain an unknown contribution. As will be seen below, the permanent dipole moment is thereby affected, but not the field-induced moment or, equivalently, the L&NLO properties (a proof for quasilinear systems is given in [2]). The same occurs for large finite systems. In that case, the contribution arises from (shape-dependent) surface charges, whereas for infinite periodic systems it has a mathematical origin. Nonetheless, they describe the same situation. In particular, for the quasilinear case, the direct connection between surface charge and the phase of the crystal orbitals has been shown quantitatively [10]. In general, however, the relationship between a given shape and surface charges of a finite system and the additive term for the corresponding infinite, periodic system has not yet been made.

In addition to the electronic dipole moment one must add the contribution from the nuclei,

$$\bar{\mu}_n = \frac{1}{K} \sum_m Z_m e \mathbf{R}_m, \tag{3.13}$$

in order to obtain the total value. In Eq. (3.13) the m summation runs over all nuclei in the BvK zone. If the origin is chosen to be at the center of nuclear charge, then this term vanishes.

As mentioned above, the presence of an electric field can be introduced through either a scalar potential or a vector potential (as well as combinations of the two). We have just seen how the dipole moment can be described for an infinite periodic system. In principle, the prescription given opens up the possibility of including the interaction with an electrostatic field in the Hamiltonian by means of the scalar potential.

This approach has been developed [11, 12] but, first, we turn to the slightly earlier treatment of Kirtman, Gu, and Bishop, known as the KGB method [1, 2], based on their vector potential approach (VPA). They considered a homogeneous (i.e. constant in space), time-dependent electric field, in which case the electronic response can be calculated from the time-dependent single-particle equation (3.1) with

$$\frac{\partial}{\partial t} \mathbf{A}_{\text{ext};E}(\mathbf{r}, t) = -c\mathbf{E}(\mathbf{r}, t) = -c\mathbf{E}(t)$$
$$\mathbf{A}_{\text{ext};E}(\mathbf{r}, t) = \mathbf{A}(t)$$
$$V_{\text{ext};E}(\mathbf{r}, t) = 0. \tag{3.14}$$

Since $\mathbf{A}$ is homogeneous its presence in the kinetic energy term does not destroy the translational symmetry. Accordingly, the (time-dependent) electronic orbitals can be written as in Eq. (3.9) but now the lattice function depends on time not only explicitly, but also implicitly through the replacement of $\mathbf{k}$ by:

$$\mathbf{k} \rightarrow \kappa = \mathbf{k} + \frac{e}{c}\mathbf{A}(t). \tag{3.15}$$

The electric field terms enter into the single particle equation (3.1) when this dependence upon κ is taken into account in obtaining the time derivative, which contains the term $\nabla_\kappa u_j(\kappa, \mathbf{r}, t) \cdot \frac{d\kappa}{dt}$ with $\frac{d\kappa}{dt}$ equal to $-e\mathbf{E}(t)$. After taking the derivative we may return from κ to $\mathbf{k}$ because either may be used to calculate any observable by integration over the complete first Brillouin zone. The only limitation in doing so is the requirement that there be a gap between occupied and unoccupied orbitals. Accordingly, the electronic wavefunctions in the resulting single particle equation can be expanded in terms of ordinary Bloch functions, i.e.

$$\psi_j(\mathbf{k}, \mathbf{r}, t) = \sum_l \chi_l(\mathbf{k}, \mathbf{r}) C_{lj}(\mathbf{k}, t), \tag{3.16}$$

with

$$\chi_l(\mathbf{k}, \mathbf{r}) = \frac{1}{\sqrt{N}} \sum_n e^{i\mathbf{k}\cdot\mathbf{R}_n/\hbar} \chi_{ln}(\mathbf{r}). \tag{3.17}$$

The Bloch function in Eq. (3.17) is constructed by combining equivalent basis functions (index l) of different unit cells (index n). It follows (see Ref. [1]) that Eq. (3.1) can be written in matrix form as

$$\underline{\underline{F}}(\mathbf{k}, t) \cdot \underline{C}(\mathbf{k}, t) + e\mathbf{E}(t) \cdot \left[\underline{\underline{M}}(\mathbf{k}) \cdot \underline{C}(\mathbf{k}, t) + i\hbar\underline{\underline{S}}(\mathbf{k})\nabla_k\underline{C}(\mathbf{k}, t)\right]$$

$$-i\hbar\underline{\underline{S}}(\mathbf{k})\frac{\partial}{\partial t}\underline{C}(\mathbf{k}, t) = \underline{C}(\mathbf{k}, t)\underline{\underline{S}}(\mathbf{k}) \cdot \underline{\underline{\varepsilon}}(\mathbf{k}, t). \tag{3.18}$$

where

$$S_{qp}(\mathbf{k}) = \sum_n e^{i\mathbf{k}\cdot\mathbf{R}_n/\hbar}\langle\chi_{q0}|\chi_{pn}\rangle$$

$$F_{qp}(\mathbf{k}, t) = \sum_n e^{i\mathbf{k}\cdot\mathbf{R}_n/\hbar}\langle\chi_{q0}|\hat{F}(t)|\chi_{pn}\rangle$$

$$M_{qp}(\mathbf{k}) = \sum_n e^{i\mathbf{k}\cdot\mathbf{R}_n/\hbar}\langle\chi_{q0}|\mathbf{r} - \mathbf{R}_n|\chi_{pn}\rangle = \sum_n e^{-i\mathbf{k}\cdot\mathbf{R}_n/\hbar}\langle\chi_{qn}|\mathbf{r} - \mathbf{R}_0|\chi_{p0}\rangle$$

$$\tag{3.19}$$

are the overlap, single particle Hamiltonian (Fock or Kohn-Sham), and unit cell dipole matrices, respectively. $\mathbf{R}_n$ is a typical position of the nth unit cell, while the operator $\hat{F}$ is the usual Fock operator discussed in Chap. 2. The latter depends implicitly on the field through the density matrix, but is otherwise field-free. Finally, the matrix of

Lagrange multipliers $\underline{\underline{\varepsilon}}(\mathbf{k}, t)$ assures the orthonormality of the orbitals. It should be emphasized that this treatment maintains $\mathbf{k}$ as a good set of quantum numbers.

A special case is that of a static electric field, whereby Eq. (3.18) takes the form

$$\left\{\underline{\underline{F}}(\mathbf{k}) + e\mathbf{E} \cdot \left[\underline{\underline{M}}(\mathbf{k}) + i\hbar\underline{\underline{S}}(\mathbf{k})\nabla_k\right]\right\} \cdot \underline{C}_j(\mathbf{k}) = \varepsilon_j(\mathbf{k}) \cdot \underline{\underline{S}}(\mathbf{k}) \cdot \underline{C}_j(\mathbf{k}). \qquad (3.20)$$

for the jth orbital (assuming the canonical, i.e. diagonal, form for the Lagrange multiplier matrix). This equation is the same as that later derived [12] using the scalar potential description of the electrostatic field and Eq. (3.12) for the electronic contribution to the dipole moment per unit. Thus, the quantity in square brackets is the effective dipole moment (per unit) operator. Equation (3.18) has an additional advantage that it immediately allows the treatment of dynamic, as well as static, fields as required for the study of L&NLO properties.

Equation (3.20) is not a simple eigenvalue problem due to the presence of the ∇_k operator. In the absence of the electrostatic field, the resulting orbital expansion coefficients $\underline{C}_j(\mathbf{k})$ for a general $\mathbf{k}$ will be complex with phase factors that, essentially, vary arbitrarily from one $\mathbf{k}$ to the next. For most purposes this does not matter. However, for non-zero $\mathbf{E}$, it becomes impossible to evaluate the effect of the ∇_k operator numerically without prior treatment of the coefficients. Moreover, when modifying the orbital expansion coefficients by introducing an arbitrary phase factor

$$\underline{C}_j(\mathbf{k}) \rightarrow e^{i\phi_j(\mathbf{k})}\underline{C}_j(\mathbf{k}), \qquad (3.21)$$

the orbital energies change as well

$$\varepsilon_j(\mathbf{k}) \rightarrow \varepsilon_j(\mathbf{k}) - e\mathbf{E} \cdot \nabla_k\phi_j(\mathbf{k}). \qquad (3.22)$$

This implies that, in the presence of an electrostatic field, the band structure becomes non-unique and ultimately loses physical meaning [13].

The coefficients $\underline{C}_j(\mathbf{k})$ can be made smooth by choosing phases that vary minimally as a function of $\mathbf{k}$. Then, application of the ∇_k operator on the coefficients becomes numerically stable [12]. In fact, numerical smoothing turns out to yield results that are more stable than an earlier analytical procedure [1, 2] according to which the derivative is obtained using the matrix $\underline{\underline{Q}}_\zeta$ defined by

$$\frac{\partial}{\partial k_\zeta}\underline{\underline{C}}^{(0)}(\mathbf{k}) = \underline{\underline{C}}^{(0)}(\mathbf{k})\underline{\underline{Q}}_\zeta(\mathbf{k}) \qquad (3.23)$$

with $\underline{\underline{Q}}_\zeta$ determined largely (but not quite entirely) by differentiation of the Fock equation. Here we have explicitly introduced the superscript $^{(0)}$ to emphasize that $\underline{\underline{C}}^{(0)}(\mathbf{k})$ is the matrix containing the field-free orbital expansion coefficients. The derivatives of the higher-order coefficients are, then, obtained in terms of $\underline{\underline{Q}}_\zeta$ and derivatives of matrices that provide the solution of the perturbation equations (see Chap. 2).

Even with smoothing, the left-hand side of Eq. (3.20) does not have the form of a matrix multiplying the unknown coefficient vector $\underline{C}_j(\mathbf{k})$. In order to obtain that form one may utilize the normalization condition.

$$\underline{C}^{\dagger}(\mathbf{k})\underline{\underline{S}}(\mathbf{k})\underline{C}(\mathbf{k}) = \underline{1}, \tag{3.24}$$

This condition is valid both in the absence and in the presence of the field. We use it here for the field-dependent coefficients, which leads to

$$\underline{\underline{S}}(\mathbf{k})\nabla_k\underline{C}(\mathbf{k}) = \left[\underline{\underline{S}}(\mathbf{k})\left\{\nabla_k\underline{C}(\mathbf{k})\right\}\underline{C}^{\dagger}(\mathbf{k})\underline{\underline{S}}(\mathbf{k})\right]\underline{C}(\mathbf{k}). \tag{3.25}$$

After this result is inserted into Eq. (3.20) the term in square brackets may be treated self-consistently just like the usual treatment of the Fock matrix [12]. In doing so $\nabla_k\underline{C}(\mathbf{k})$ is calculated by means of the above-mentioned smoothing procedure combined with an accurate numerical differentiation (see [12]).

The self-consistent procedure is continued until the coefficients do not change any more (within a pre-chosen accuracy). For a non-vanishing $\mathbf{E}$, however, it is not possible to bring the calculation to convergence with arbitrarily high accuracy. Instead, the difference between the output and input coefficients reaches a minimum (that is larger for larger $\mathbf{E}$) and subsequently increases to some finite value. This non-convergence is related to oscillations in the coefficients $\underline{C}_j(\mathbf{k})$ and their phases. It may very well be related to the tunneling behavior that was mentioned briefly above since an analogous situation occurs for long finite chains. The point is that it limits the range of field strengths that can be considered while requiring a tolerant convergence criterion as well as carefully chosen methods for mixing output and input coefficient vectors.

For an electrostatic field one may use Eq. (3.20) to obtain an expression for the total electronic energy (per unit). Because of the ∇_k operator this expression is more complex than for an ordinary finite non-periodic system. However, it can be evaluated, and analytical derivatives with respect to structural coordinates can be obtained [12]. Thus, it is possible to carry out automatic structure optimizations in the presence of the field and, thereby, to calculate vibrational L&NLO properties by the FF-NR procedure described in Chap. 2.

Previously, we mentioned that the electronic contribution to the dipole moment per unit, $\bar{\mu}_e$, is not unique and that different values can result upon changing the phases of the electronic orbitals as in Eq. (3.21). Through the smoothing procedure, a specific value will result and other values will, in general, require phase discontinuities in $\mathbf{k}$ space. Again, quasi-one-dimensional (quasi-1D) systems are an exception. In that case, different values can occur without introducing phase discontinuities. These values differ from one another by an integer multiple (n) of the lattice constant a (times the elementary charge e). As discussed earlier, n is related to the charge at the terminations of the quasi-1D system (i.e. the surface charge). In fact, it corresponds to the transfer of n units of electronic charge from one end of the system to the other. Since the resulting change in $\bar{\mu}_e$ is proportional to a this dipole term will contribute to the piezoelectric effect. It will also alter the equilibrium geometry through coupling in the potential energy between the lattice parameter and structural coordinates of the unit cell. This coupling will also affect the vibrational L&NLO properties. For large finite systems an analogous situation occurs, although it is more complicated because the transfer of charge is non-integral and depends upon the shape of the system. Model calculations for the quasi-1D case [10] indicate that charge transfer effects can be significant, at least as far as piezoelectricity is concerned. This conclusion is supported by subsequent estimates based on density functional calculations carried out for a couple of crystalline layered perovskites [14].

3.5 TDHF and TDDFT Equations for Infinite Periodic Systems

As in the molecular case, discussed in Sect. 2.4, a coupled-perturbed treatment can be developed for calculating the clamped nucleus static and dynamic electronic responses of an infinite periodic system to electric fields. Using the VPA, the perturbation operator retains translational symmetry so that $\mathbf{k}$ remains a good quantum vector [see Eq. (3.18)]. There are only two significant differences with respect to the molecular problem: (i) the matrix equations become $\mathbf{k}$ dependent and, accordingly, must be solved separately for each $\mathbf{k}$; and (ii) the perturbation involves an operator, ∇_k, that acts on the orbital expansion coefficients.

Taking those changes into account the expressions in Tables 2.1, 2.2, and 2.3 can be converted to the infinite periodic case simply by making the correspondences (again, as in Sect. 2.4, omitting the upper indices and letting the number of lower indices give the order):

$$F_\zeta \rightarrow \Theta_\zeta + e \left[M_\zeta + i S \frac{\partial C}{\partial k_\zeta} C^\dagger S \right]$$

$$F_{\zeta\eta} \rightarrow \Theta_{\zeta\eta} + ei S C \frac{\partial U_\eta}{\partial k_\zeta} C^\dagger \qquad (3.26)$$

where Θ is due to the difference between $F(t)$ and $F^{(0)}$ in Eq. (3.19) and U_η is the first-order correction (in the field strength) to the U matrices introduced in Sect. 2.4. In the first line one may utilize Eq. (3.23) to obtain

$$\frac{\partial C}{\partial k_\zeta} C^\dagger = C Q_\zeta C^\dagger. \qquad (3.27)$$

and in deriving the second line we have used

$$
\begin{aligned}
\frac{\partial C_\eta}{\partial k_\zeta} C^\dagger + \frac{\partial C}{\partial k_\zeta} C_\eta{}^\dagger &= \frac{\partial}{\partial k_\zeta} \left[C U_\eta \right] C^\dagger + C Q_\zeta U_\eta{}^\dagger C^\dagger \\
&= \frac{\partial C}{\partial k_\zeta} U_\eta C^\dagger + C \frac{\partial U_\eta}{\partial k_\zeta} C^\dagger + C Q_\zeta U_\eta{}^\dagger C^\dagger \\
&= C \left\{ Q_\zeta \left[U_\eta + U_\eta{}^\dagger \right] + \frac{\partial U_\eta}{\partial k_\zeta} \right\} C^\dagger \\
&= C \frac{\partial U_\eta}{\partial k_\zeta} C^\dagger
\end{aligned}
\tag{3.28}
$$

since U_η is anti-hermitean. Of course, each of the corresponding (infinite periodic) quantities in Eq. (3.26) is now $\mathbf{k}$-dependent. Just like the analogous term in the molecular case Θ is determined by the field-induced change in the density matrix with the indices denoting the field direction(s).

The TDHF approach for infinite periodic systems based on the vector potential was originally developed for application to L&NLO of polymers [2]. An informative example, considered in [2], is the prototypical case of polyacetylene. This original treatment was later generalized to 2- and 3-dimensional systems, specifically for static hyperpolarizabilities, by Ferrero et al. [15]. The extension to static Kohn-Sham DFT was subsequently carried out using the same prescription as in Chap. 2 (a specific example is given in Ref. [17] of that chapter). Finally, both methods are now available in the most recent version of the CRYSTAL code (CRYSTAL14), which includes frequency-dependent polarizability, but not hyperpolarizabilities. It is expected that the latter will be implemented in the not-too-distant future.

References

1. Kirtman, B., Gu, F.L., Bishop, D.M.: Extension of the Genkin and Mednis treatment for dynamic polarizabilities and hyperpolaribilities of infinite periodic systems. I. Coupled perturbed Hartree-Fock theory. J. Chem. Phys. **113**, 1294–1309 (2000)
2. Bishop, D.M., Gu, F.L., Kirtman, B.: Coupled-perturbed Hartree-Fock theory for infinite periodic systems: calculation of static electric properties of $(LiH)_n$, $(FH)_n$, $(H_2O)_n$, $(-CNH-)_n$, and $(-CH=CH-)_n$. J. Chem. Phys. **114**, 7633–7643 (2001)
3. Vanderbilt, D., King-Smith, R.D.: Electric polarization as a bulk quantity and its relation to surface charge. Phys. Rev. B **48**, 4442–4455 (1993)
4. Kudin, K.N., Car, R., Resta, R.: Quantization of the dipole moment and of the end charges in push-pull polymers. J. Chem. Phys. **127**, 194902 (2007)
5. Springborg, M., Kirtman, B.: How much can donor/acceptor-substitution change the responses of long push-pull systems to DC fields? Chem. Phys. Lett. **454**, 105–113 (2008)
6. Blount, E.I.: Formalisms of band theory. Solid State Phys. **13**, 305–373 (1962)
7. Resta, R.: Macroscopic polarization in crystalline dielectrics: the geometric phase approach. Rev. Mod. Phys. **66**, 899–915 (1994)
8. King-Smith, R.D., Vanderbilt, D.: Theory of polarization of crystalline solids. Phys. Rev. B **47**, 1651–1654 (1993)

9. Springborg, M., Kirtman, B., Dong, Y.: Electronic polarization in quasilinear chains. Chem. Phys. Lett. **396**, 404–409 (2004)
10. Springborg, M., Tevekeliyska, V., Kirtman, B.: Termination effects in electric field polarization of periodic quasi-one-dimensional systems. Phys. Rev. B **82**, 165442 (2010)
11. Nunes, R.W., Gonze, X.: Berry-phase treatment of the homogeneous electric field pertubation in insulators. Phys. Rev. B **63**, 155107 (2001)
12. Springborg, M., Kirtman, B.: Analysis of vector potential approach for calculating linear and nonlinear responses of infinite periodic systems to a finite static external electric field. Phys. Rev. B **77**, 209901 (2008). (Phys. Rev. B **77**, 045102 (2008). Erratum)
13. Vargas, J., Springborg, M.: Building band structures for long finite chains in presence of an electric field. J. Chem. Phys. **137**, 144108 (2012)
14. Sayede, A., Bruyer, E., Springborg, M.: Ab initio study of metastable layered perovskites $R_2Ti_2O_7$ (R = Sm and Gd). Phys. Rev. B **86**, 125136 (2012)
15. Ferrero, M., Rérat, M., Kirtman, B., Dovesi, R.: Calculation of first and second static hyperpolarizabilities of one- to three-dimensional periodic compounds. Implementation in the CRYSTAL code. J. Chem. Phys. **129**, 244110 (2008)

Chapter 4
The Elongation Method

Abstract The elongation (ELG) method is a theoretical procedure for building up an arbitrary system by adding small fragments, one by one, to an original and growing cluster. In this chapter the basic features of this method are elaborated by considering, first, the field-free problem at the single particle level. Of particular importance are the procedures for regional molecular orbital localization, as well as integral (and other) cut-offs, that act in concert to allow calculations to be restricted to an interactive region of essentially fixed size. In carrying out calculations there will sometimes be a limited number of delocalized molecular orbitals that cannot meet the localization criteria. A generalized technique for dealing with this circumstance, as well as two- and three-dimensional systems is described. The predicted overall linear scaling behavior is shown to be verified in practice. Finally, building upon the single particle case, we present our formulation of ELG-LMP2 for electron correlation and ELG-LCIS for excited electronic states.

4.1 Introduction

Conventional ab initio quantum-chemical calculations are typically based on canonical molecular orbitals (CMOs) that are extended over the entire system. The CMOs are, in fact, just one particular solution of the Hartree-Fock, or Kohn-Sham, equations for which the matrix of Lagrange multipliers is diagonal. In fact, the Hartree-Fock (or Kohn-Sham) wavefunctions are unaltered by any unitary transformation of the occupied and the vacant MOs if they are not mixed. Thus, any other set of orthonormal MOs, such as localized MOs, can equally well be used.

The disadvantage of the CMOs is that they do not directly relate to most traditional chemical concepts, and it is not possible to utilize the locality of the underlying electronic structure. Consequently, CMO-based calculations require information about the entire system including a large number of two-electron integrals that can lead to computational difficulties even for intermediate sized molecules. Although small molecules can be treated by sophisticated wavefunction methods, and the results obtained can even sometimes be used to calibrate experimental measurements, CMO-based quantum chemical approaches are not easily applied to large systems, especially at the ab initio level.

© The Author(s) 2015
43
F.L. Gu et al., *Calculations on nonlinear optical properties for large systems*,
SpringerBriefs in Electrical and Magnetic Properties of Atoms,
Molecules, and Clusters, DOI 10.1007/978-3-319-11068-4_4

If there is translational symmetry, one can apply periodic boundary conditions (PBC) [1], as discussed in the previous chapter to simplify the calculations. Indeed, remarkable progress has been made in computing the electronic structure of polymers [2–4], surfaces and crystals [5, 6] that are periodic in 1D, 2D, or 3D respectively. The elongation method is particularly potent in cases where the system of interest does not possess translational symmetry.

In this chapter we give a brief introduction to the elongation method [7–13] for the field-free case. In order to outline the method we use a random polymer chain for illustrative purposes. This chain is constructed by adding monomer units one by one to a starting cluster, as illustrated in Fig. 4.1. Our procedure mimics the experimental polymerization/copolymerization synthetic mechanism. The computational advantages originate from the fact that before each addition, a set of localized molecular orbitals (LMOs) is constructed and the subset of LMOs localized on regions far from the reaction site, is kept frozen. This approach allows the number of variational degrees of freedom in the ELG-SCF procedure to be kept practically constant.

Our Hartree-Fock (Kohn-Sham) regional localization scheme is presented in Sect. 4.2. Given an appropriate starting cluster the first step is to solve the SCF equation for that cluster. Next, the system is divided into two parts. One part, the

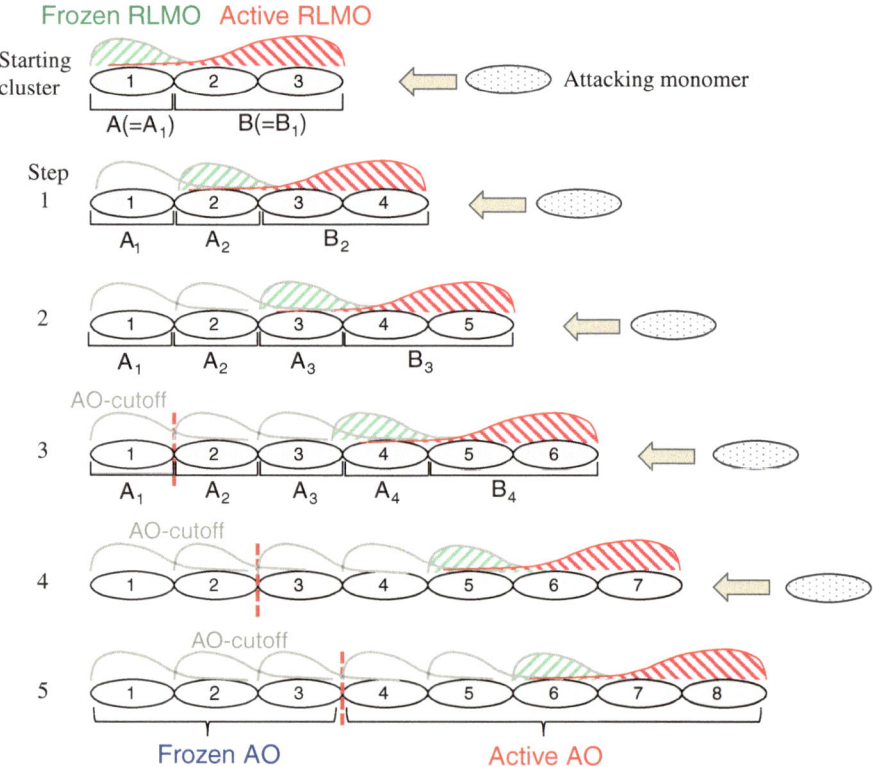

Fig. 4.1 Schematic illustration of the elongation method

frozen region, consists of atomic orbitals (AOs) that are far away from the chain propagation site; while the other part, the active region, is at the chain propagation site. Then the CMOs are transformed to regionally localized molecular orbitals (RL-MOs) for the two regions by a localization procedure especially designed for that purpose [14].

Once the RLMOs have been obtained we can proceed to the elongation step, described in Sect. 4.3, in which a monomer unit is added to the chain. The active region is enlarged to include the monomer AOs and the resulting Hartree-Fock (Kohn-Sham) equation is solved self-consistently (ELG-SCF). Subsequently, a new active region is defined so that it is essentially the same in size as the original active region and the frozen region is enlarged accordingly. After that is done the system is ready for the next ELG-SCF step. This process is repeated by adding one monomer at a time until an overall system size is reached for which the quantity of interest is converged.

It is important to note that the ELG-SCF calculation is limited to the active space. This leads to a substantial reduction in computational costs. The saving is of major significance for very large systems where diagonalization with its $O(N^3)$ scaling turns out to be the computational bottleneck. Nevertheless, to realize the full potential of the method one must also reduce the computational cost of Fock matrix construction. To that end we use two tools and the combination of them. One tool is the standard quantum fast multipole method combined with linear-scaling exchange matrix construction algorithm [15–20]. The second is the cut-off technique of Sect. 4.4 which allows us to avoid calculating a large part of the Fock matrix during the SCF process [21, 22]. The cut-off technique does not further reduce the already near linear scaling of the procedure, but has been shown to significantly reduce its prefactor.

The elongation method has been successfully used in a variety of applications including calculations of band structure [23–25], electronic structure of biomolecules [26] and, of primary interest here, nonlinear optical properties [27–34]. Some results for quasi-one-dimensional (quasi-1D) systems with large unit cells, that illustrate linear scaling, are shown in Sect. 4.5. Moreover, the original method has now been generalized for treating certain problematic quasi-1D cases as well as two- and three-dimensional systems. Section 4.6 outlines the approach used in the generalized ELG (G-ELG) method [35–38] and in Sect. 4.7 some applications including solvent effect are presented. Finally, the Hartree-Fock wavefunction treatment has been extended to include electron correlation at the MP2 level [39] (see Sect. 4.8) and CI-singles for excited states [40] (see Sect. 4.9).

4.2 Localization Scheme for the Elongation Method

The localization scheme is a critical part of the elongation method. It is carried out prior to each elongation step in order to move localized orbitals from the active space into the frozen orbital region. The newly frozen orbitals will be designated region

A in the following, while the remainder of the active cluster orbitals, used in the subsequent ELG-SCF calculation, constitutes region B.

Initially, the cluster must be large enough so that the localization scheme produces a set of well-localized orbitals for a pre-defined (by the user) fragment (=region). This is fragment 1 shown in the starting cluster of schematic Fig. 4.1—it is also region $A = A_1$. The remainder, shown as fragments 2 (see later) and 3 constitutes region $B = B_1$, which is the active space of the cluster. Choosing an appropriate size for the initial cluster may require some trial and error, although enough experience has now been gained to make a good first choice. The approach for testing whether the orbitals of A_1 are sufficiently well-localized, with insignificant tails in B_1, is described at the beginning of Sect. 4.6.

After a (pre-defined) attacking fragment (=fragment 4) has been added to the initial cluster in the ELG-SCF procedure, we seek to extend the frozen region so as to include fragment 2. Hence, the localization in step 1 is carried out on that part of the cluster composed of fragments 2, 3 and 4 with fragment 2 equal to $A = A_2$, while fragments 3 and 4 constitute $B = B_2$. Again, a test must be carried out to ensure that the orbitals of A_2 are adequately localized in order to include them as part of the total frozen region, i.e. the sum of A_1 and A_2. If not, then B_2 is enlarged to include fragment 2 in the active space along with fragments 3 and 4. In either event, the next part is to add another fragment (=fragment 5) through an ELG-SCF calculation (see the next section). Subsequent steps repeat this general procedure whereby the frozen region gradually becomes larger.

We need to discuss how the regionally localized molecular orbitals (RLMOs) are obtained. As already mentioned, the CMOs of the starting cluster, which are found by a conventional quantum chemistry calculation, are naturally delocalized over the entire active space. They have the following LCAO form:

$$\underline{\varphi}^{CMO} = \underline{\chi}^{AO} \underline{\underline{C}}^{CMO}_{AO} \tag{4.1}$$

where $\underline{\varphi}^{CMO}$ is a row vector of atomic orbitals (AOs) in the basis set for the starting cluster, and $\underline{\underline{C}}^{CMO}_{AO}$ is the (AO × CMO) coefficient matrix obtained by solving the Hartree-Fock (HF) or Kohn-Sham DFT (KS-DFT) equation

$$\underline{\underline{F}}\,\underline{\underline{C}} = \underline{\underline{S}}\,\underline{\underline{C}}\,\underline{\underline{\varepsilon}} \tag{4.2}$$

Here $\underline{\underline{S}}$ and $\underline{\underline{F}}$ are, respectively, the AO overlap and effective one-electron Hamiltonian matrices.

In general for steps 1, 2, ... the AOs of the cluster are replaced by the RLMOs of the active region B plus the AOs of the attacking fragment. The localization procedure is exactly the same in either case. Hence, we will continue to use the same notation as for the starting cluster. In addition, we specialize below to the HF method. Although the KS-DFT treatment is not exactly the same (cf. Chap. 2), the differences are small enough that the reader will easily be able to adapt from the former to the latter.

The delocalized CMOs of the starting cluster (or those obtained in the ELG-SCF procedure) are localized into two sets corresponding to a frozen region and an active region as discussed above. An efficient and reliable scheme for obtaining well-localized RLMOs has been developed by Gu et al. [14]. Their scheme starts by converting from AOs to orthonormalized AOs (OAOs) using Löwdin's symmetric orthogonalization procedure [41]. Given that the original coefficient matrix satisfies ($\underline{\underline{1}}$ is the identity matrix):

$$\underline{\underline{C}}_{AO}^{CMO\dagger} \, \underline{\underline{S}}^{AO} \, \underline{\underline{C}}_{AO}^{CMO} = \underline{\underline{1}} \tag{4.3}$$

the OAOs may be written as

$$\underline{\underline{C}}_{OAO}^{CMO} = \underline{\underline{X}} \, \underline{\underline{C}}_{AO}^{CMO} \tag{4.4}$$

where

$$\underline{\underline{X}} = \underline{\underline{V}} \, \underline{\underline{\lambda}}^{1/2} \, \underline{\underline{V}}^{\dagger} = \underline{\underline{X}}^{\dagger} \tag{4.5}$$

and $\underline{\underline{V}}$ is the unitary matrix that diagonalizes $\underline{\underline{S}}^{AO}$, i.e.

$$\underline{\underline{S}}^{AO} \, \underline{\underline{V}} = \underline{\underline{V}} \, \underline{\underline{\lambda}}. \tag{4.6}$$

Löwdin's transformation minimizes the distortion of the OAO basis functions with respect to the original AOs. At the same time, the density matrix in the AO basis (see Chap. 2), defined as:

$$\underline{\underline{D}}^{AO} = \underline{\underline{C}}_{AO}^{CMO} \, \underline{\underline{n}} \, \underline{\underline{C}}_{AO}^{CMO\dagger}, \tag{4.7}$$

in which $\underline{\underline{n}}$ is the diagonal occupation number matrix, becomes

$$\underline{\underline{D}}^{OAO} = \underline{\underline{X}} \, \underline{\underline{D}}^{AO} \, \underline{\underline{X}}^{\dagger}. \tag{4.8}$$

In the OAO basis the idempotency relation, $\underline{\underline{D}}^{OAO} \, \underline{\underline{D}}^{OAO} = 2\underline{\underline{D}}^{OAO}$, guarantees that the eigenvalues of $\underline{\underline{D}}^{OAO}$ are either 0 or 2 (assuming spin-restricted HF) which means that the eigenvectors will correspond to orbitals that are either completely occupied or completely unoccupied as desired. However, diagonalization will inevitably spoil the localization. In order to make that effect as small as possible, the desired RLMOs are obtained in two steps.

In the first step regional orbitals (ROs) that are localized, but not fully occupied, are constructed by separately diagonalizing the sub-blocks of $\underline{\underline{D}}^{OAO}$ associated with AOs belonging to either frozen region A or active region B. Thus, the transformation from OAOs to ROs is given by the direct sum of $\underline{\underline{T}}^A$ and $\underline{\underline{T}}^B$, i.e.:

$$\underline{\underline{T}} = \underline{\underline{T}}^A \oplus \underline{\underline{T}}^B \tag{4.9}$$

where $\underline{T}^A$ and $\underline{T}^B$ are the eigenvectors of $\underline{\underline{D}}^{OAO}(A)$ and $\underline{\underline{D}}^{OAO}(B)$, respectively. These ROs are partially occupied and partially unoccupied. They can be divided into three sets that are either approximately doubly-occupied (eigenvalue is close to 2), approximately empty (eigenvalue close to 0), or approximately singly-occupied (eigenvalue is close to 1). Each singly occupied orbital in A is the hybridized atomic component of a bond orbital that can be combined with the corresponding atomic component in B to form a covalent bonding or antibonding pair. Instead of doing that, however, we transfer an electron from each singly occupied orbital of A to the corresponding singly occupied orbital of B or vice versa. This does not affect the T matrix in Eq. (4.9), but does determine which ROs are approximately doubly occupied and which are approximately empty by altering $\underline{\underline{D}}^{OAO}$ in Eq. (4.11) below. It is unnecessary for a non-bonded system, such as a water chain (which we will consider later for test purposes), since there are no singly occupied orbitals in that case. After transformation to ROs, the resulting coefficient and density matrices are

$$\underline{\underline{C}}_{RO}^{CMO} = \underline{\underline{T}}^{\dagger} \underline{\underline{C}}_{OAO}^{CMO} \tag{4.10}$$

and

$$\underline{\underline{D}}^{RO} = \underline{\underline{T}}^{\dagger} \underline{\underline{D}}^{OAO} \underline{\underline{T}} \tag{4.11}$$

Using Eq. (4.11), the unitary condition $\underline{\underline{T}} \, \underline{\underline{T}}^{\dagger} = \underline{\underline{T}}^{\dagger} \underline{\underline{T}} = \underline{\underline{1}}$, and the idempotency relation for $\underline{\underline{D}}^{OAO}$, one can easily verify that

$$\underline{\underline{D}}^{RO} \underline{\underline{D}}^{RO} = 2\underline{\underline{D}}^{RO}. \tag{4.12}$$

In the second step, we perform a unitary transformation between the occupied and unoccupied blocks of $\underline{\underline{D}}^{RO}$ in order to diagonalize this matrix, but in such a way as to preserve the localization as much as possible. The procedure is the same as the construction of natural bond orbitals (NBOs) [42] except that it is generalized in this context to localized regional orbitals. If $\underline{\underline{U}}$ is the unitary matrix so constructed, i.e.

$$\underline{\underline{D}}^{RLMO} = \underline{\underline{U}}^{\dagger} \underline{\underline{D}}^{RO} \underline{\underline{U}}, \tag{4.13}$$

then

$$\underline{\underline{C}}_{RLMO}^{CMO} = \underline{\underline{U}}^{\dagger} \underline{\underline{C}}_{RO}^{CMO} \quad \text{or} \quad \underline{\underline{C}}_{AO}^{RLMO} = \underline{\underline{X}}^{-1} \underline{\underline{T}} \, \underline{\underline{U}} \tag{4.14}$$

where the transformation from AOs to RLMOs is obtained by utilizing Eqs. (4.4) and (4.10).

The localization scheme just described is more efficient and accurate than earlier versions of the elongation method. It has been implemented and linked to the GAMESS program package in which elongation calculations of electronic structures at the HF, MP2, and DFT levels are now available. A variety of tests on model 1D systems ranging from non-bonded molecular chains to highly delocalized polymers show that, with this scheme, the elongation error per unit cell is satisfactorily small (less than 10^{-6} a.u.) as long as the starting cluster is sufficiently large—typically five units for non-bonded systems and ten units for bonded systems. Our experience, thus far, indicates that the same is true for 2D and 3D structures.

4.3 ELG-SCF

In Fig. 4.1 the elongation is carried out on the sub-space that consists of the active region B on the cluster plus the attacking monomer (M). For the starting cluster B consists of fragments 2 plus 3; for step 1 it consists of fragments 3 + 4; etc. After localization to get RLMOs as described in the previous section we write the orthonormal RLMOs for the separate regions of the cluster, in the general case, as:

$$\varphi_i^{RLMO}(X) = \sum_{j=1}^{N} \sum_{\mu} L_{\mu i}^{(j)}(X) \chi_{\mu}^{(j)} \tag{4.15}$$

where the superscript j is the index for the jth fragment, $L_{\mu i}^{(j)}(X)(X = A \text{ or } B)$ is the RLMO coefficient, and A is the entire frozen orbital region. Note that the double sum includes all AOs of the entire system, that are assigned to one fragment or another.

In preparation for elongation we replace B by the combined active region B + M. The basis set for that region is taken to be the RLMOs of Eq. (4.15) plus the CMOs of M. In fact, it is not necessary to use the CMOs of M. For greater efficiency we employ an approximate CMO basis set that, in addition to being internally orthonormal, is also orthonormal to the RLMOs of B. If $\underline{\underline{C}}(M)$ is the coefficient matrix that gives these orbitals as a linear combination of the AOs on B + M, then for the combined region, we have

$$\underline{\underline{L}}^{(j)'}(B + M) = \underline{\underline{L}}^{(j)}(B) \oplus \underline{\underline{C}}(M). \tag{4.16}$$

Although the orbitals of region A are frozen they do contribute to the effective potential that is felt by the electrons in the active space. Thus,

$$F_{ij}^{RLMO-CMO}(B + M)$$
$$= \sum_{k=1}^{N} \sum_{l=1}^{N} \sum_{\mu}^{\mu_k} \sum_{\nu}^{\nu_l} \{L_{\mu i}^{\dagger(k)'}(B + M) F_{\mu\nu}^{AO}(A + B + M) L_{\nu j}^{(l)'}(B + M)\} \tag{4.17}$$

where the superscript RLMO-CMO refers to the basis set given by Eq. (4.16). In evaluating $F_{ij}^{RLMO-CMO}$ an integral cut-off procedure, described in the next section, is applied to reduce the computation time associated with summing over the entire frozen region. Given the effective Fock matrix for the active space we can proceed to solve the canonical Hartree-Fock (Kohn-Sham) equation

$$\underline{\underline{F}}^{\text{RLMO-CMO}}(B + M) \underline{\underline{V}}(B + M) = \underline{\underline{V}}(B + M) \underline{\underline{\varepsilon}}(B + M) \qquad (4.18)$$

After solving Eq. (4.18), the CMOs of the $B + M$ region are given by the overall transformation:

$$C_{\mu i}^{(j)}(B + M) = \sum_p L_{\mu p}^{(j)'}(B + M) V_{pi}(B + M) \qquad (4.19)$$

The total density matrix can, then, be constructed as:

$$D_{\mu\nu}^{\text{AO}} = \sum_j \sum_i [L_{\mu i}^{(j)}(A) n L_{\nu i}^{(j)}(A) + C_{\mu i}^{(j)}(B + M) n C_{\nu i}^{(j)}(B + M)] \qquad (4.20)$$

which leads to the total Fock matrix:

$$F_{\mu\nu}^{\text{AO}}(A + B + M) = H_{\mu\nu}^{\text{core}} + \sum_{\lambda\sigma} D_{\lambda\sigma}^{\text{AO}}[(\mu\nu|\sigma\lambda) - \frac{1}{2}(\mu\lambda|\sigma\nu)] \qquad (4.21)$$

and the total energy of the elongated system:

$$E_{\text{tot}} = \frac{1}{2} Tr\{\underline{\underline{D}} \cdot [\underline{\underline{H}}^{\text{core}} + \underline{\underline{F}}]\} + E_{\text{N-N}}. \qquad (4.22)$$

As the system is elongated, the AOs in the frozen region that are furthest remote from the active space will contribute less and less to subsequent elongation steps. That is the rationale for the cutoff technique used to eliminate some of the AOs in Eq. (4.17), the second term on the rhs of Eq. (4.20), and many of the integrals in Eq. (4.21).

4.4 ELG-CUTOFF Method

As the system is elongated, earlier frozen regions recede from the newly active region and interact more weakly with the latter. This suggests the application of a cutoff technique wherein some orbitals are removed in constructing the Fock matrix for the elongation procedure. In our AO-cutoff procedure we eliminate appropriate AOs from the AO Fock matrix and from the basis functions of active RLMOs as well. This AO-cutoff technique substantially reduces the number of two-electron integrals that are computed in a given step. Most importantly, it reduces the computation time significantly. In practice, a threshold value for the Fock matrix elements between the RLMOs of the frozen and active regions is set (normally 10^{-5}). If the interaction matrix element is smaller than the threshold, then the frozen RLMOs are disregarded in the next elongation step. In Fig. 4.1 we can see that cut-offs occur for the first time in the third step. For non-bonded systems, such as water chains, this is typical.

However, for covalently bonded systems, more steps are needed before cutoffs can be introduced (as will be seen later when we discuss Fig. 4.3).

4.5 Results for Quasi-One-Dimensional Systems with Large Unit Cells

We present here representative results of some calculations with the elongation method for quasi-one-dimensional systems with large unit cells. One example is poly [(9,9-di-n-octylfluorenyl-2,7-diyl)-alt-(benzo[2,1,3]thiadiazol-4,8-diyl)] (F8BT). The unit cell and overall structure of F8BT are given in Fig. 4.2. The first molecule in each unit is (9,9-di-n-octylfluorenyl-2,7-diyl), the second molecule is alt-(benzo[2,1,3] thiadiazol-4,8-diyl), and the attacking monomer (M) can be either an 'odd' or an 'even' unit. For this system, we elongate until 40 units are included. It is found that the error per atom is around 10^{-9} a.u., which shows the high accuracy of the elongation method.

As far as efficiency is concerned, Fig. 4.3 shows a comparison of CPU time between the elongation and conventional methods. One can see that the elongation shows linear scaling and is much more efficient than a conventional calculation for the larger systems.

Elongation calculations were also performed for two other examples: poly(9,9-n-dihexyl-2,7-fluorene-alt-9-phenyl-3,6-carbazole) (F6PC) and poly[(9,9-dihexyl-fluorene-2,7-diyl)-co-(anthracene-9,10-diyl)] (F6A), which are shown in Figs. 4.4 and 4.5, respectively. As in the case of F8BT, high accuracy and efficiency is obtained for these two examples in elongation calculations. This indicates that the elongation method can successfully treat large-unit-cell systems.

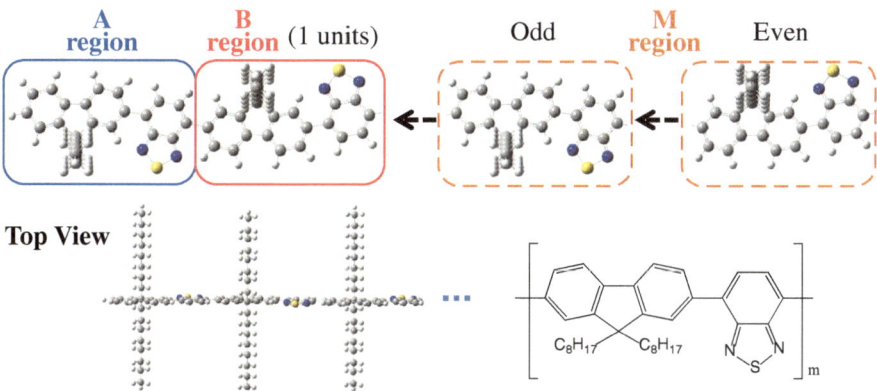

Fig. 4.2 The unit cell and overall structure of F8BT

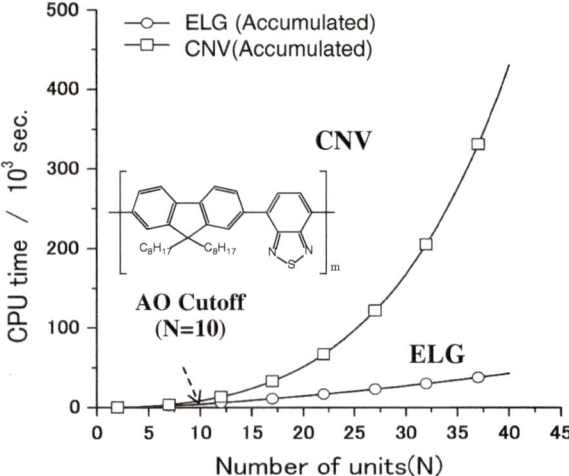

Fig. 4.3 CPU comparison between elongation (ELG) and conventional (CNV) calculations. For the elongation calculation, the AO cutoff starts at N = 10. The CPU time for both elongation and conventional calculations is the sum for all the steps, indicated as 'accumulated'

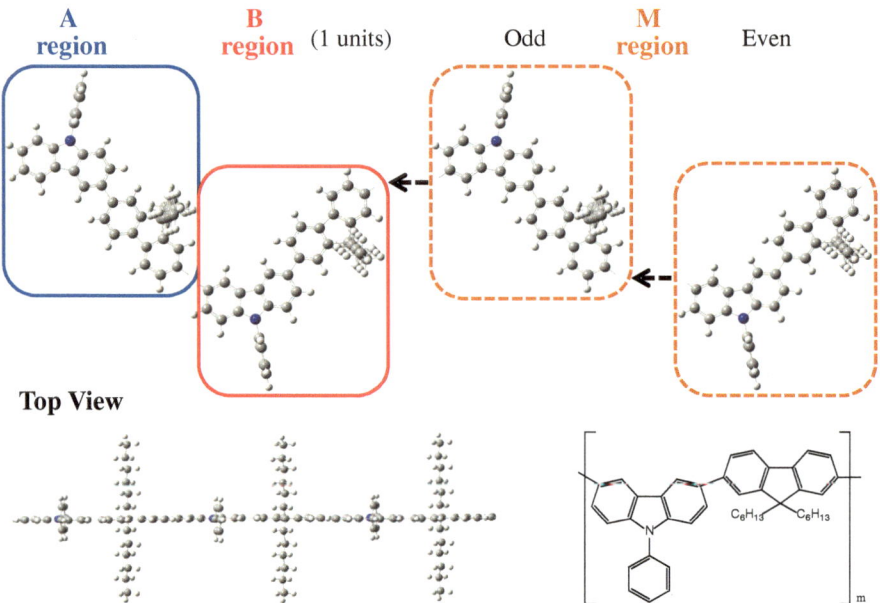

Fig. 4.4 Unit cell and overall structure of F6PC

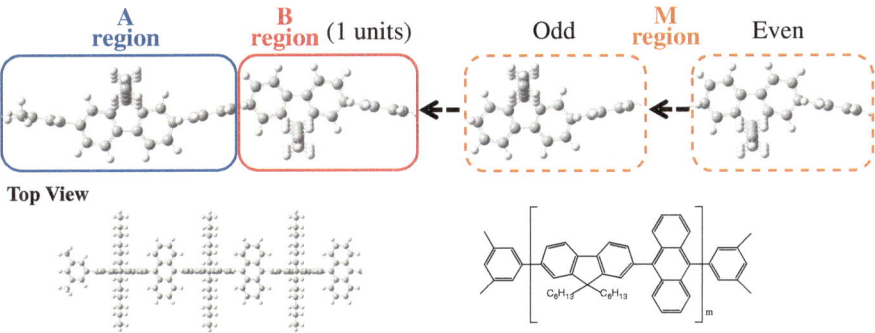

Fig. 4.5 Unit cell and overall structure of F6A

4.6 Generalized Elongation (G-ELG) Method

The frozen RLMOs have tails in the active region (see Fig. 4.1). In order to check the magnitude of these tails it is convenient to use the quantities $P_i(B)$

$$P_i(B) = \sum_r^{\text{on } B} \sum_s^{\text{on } B} |C_{ri}^{\text{frozen RLMO}} S_{rs} C_{si}^{\text{frozen RLMO}}| \qquad (4.23)$$

If $P_i(B)$ is less than a threshold value (for example, 10^{-5}), we continue to treat the orbital as a frozen RLMO. If it is not, then this RLMO is included instead in the active region. The threshold value is an important parameter. If it is set too tight, the efficiency of the elongation method will be reduced. On the other hand, if the threshold value is set too loose, the treatment will become too inaccurate. For strongly delocalized systems it turns out that there are often some RLMOs that are taken to be frozen according to the above criterion, but need to be included in the active space. Even when the system has few such orbitals, without modification the elongation method will fail to provide satisfactory accuracy for the total energy. This can cause a large error, for example, in calculating the second hyperpolarizability, which is determined by fourth-order derivatives of the energy.

We have reported [43] an orbital-shift procedure wherein strongly delocalized orbitals, i.e. orbitals that cannot be well-localized by the treatment described in Sect. 4.2, are always shifted to the active region. However, that treatment abandons the regional concept and considers the orbitals individually. This means that the regional elongation AO-cutoff technique cannot be utilized because some basis functions are always active making it impossible to freeze the region in which they are located.

Fortunately, it is possible to retain the regional concept by modifying the size and shape of the regions so that the AO-cutoff scheme becomes applicable after several steps of the elongation process. This is the basis of the generalized elongation method (G-ELG) for one-dimensional systems as shown in the upper part of Fig. 4.6. It is generally applicable unless the system is metallic (zero HOMO-LUMO gap), in

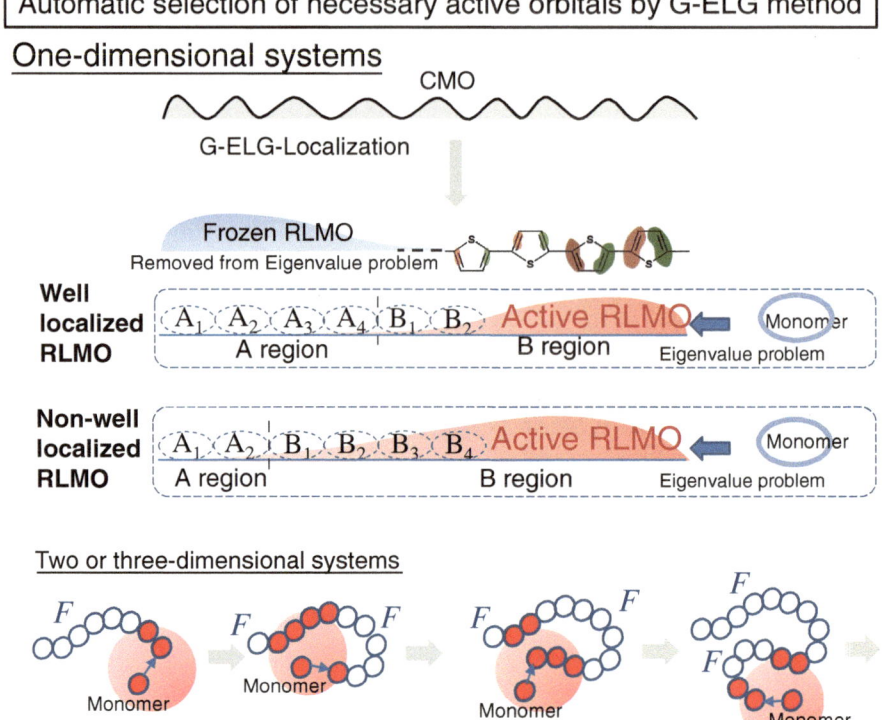

| Automatic selection of necessary active orbitals by G-ELG method |

One-dimensional systems

CMO

G-ELG-Localization

Frozen RLMO
Removed from Eigenvalue problem

Well localized RLMO
A_1 A_2 A_3 A_4 | B_1 B_2 Active RLMO ← Monomer
A region | B region   Eigenvalue problem

Non-well localized RLMO
A_1 A_2 | B_1 B_2 B_3 B_4 Active RLMO ← Monomer
A region | B region   Eigenvalue problem

Two or three-dimensional systems

F   Monomer   F   Monomer   F   Monomer   F   Monomer

Fig. 4.6 Schematic illustration of the generalized elongation (G-ELG) method

which case there is at least one orbital that cannot be localized. In that event, neither the conventional Hartree-Fock (KS-DFT) nor the G-ELG SCF can be carried out. Although in some instances the active region may have to be enlarged to account for strong delocalization, the regional-based G-ELG method with AO-cutoff preserves linear scaling.

Beyond one-dimensional systems, G-ELG will handle cases in two- or three-dimensional systems where a significant interaction between one or more frozen RLMOs and the CMOs of the attacking monomer may newly appear in an elongation step. This situation is detected automatically—when it occurs all the RLMOs of the previously frozen region(s) are treated as active RLMOs and included in the ELG-SCF procedure for addition of the attacking monomer.

Subsequently, the region that was activated can be re-frozen for the next elongation step. A schematic illustration of the G-ELG method for two- and three-dimensional systems is shown in the lower part of Fig. 4.6. The regions within the red circle are active whereas the other regions, indicated by empty blue circles (and marked 'F'), are frozen [35]. This treatment is also suitable for application of the regional AO-cutoff technique [21, 22].

The G-ELG method includes the possibility of carrying out elongation simultaneously in several directions, for example at both terminals of a one-dimensional system or along different branches in two- or three-dimensional dendrimer-like systems. The CMOs of a starting cluster are localized into a central A region and various terminal B regions. This treatment has great advantage for investigating the effect of different terminal attachments to a common wire. One first does a calculation for the central wire and then, after saving the RLMOs, the wire is terminated with various combinations of donor-acceptor species.

4.7 Some Applications of the G-ELG Method

4.7.1 Cyclic Array of Meso-Meso Linked Porphyrins

The G-ELG treatment has been successfully tested [35–38], for example, on entangled biomolecules and on intramolecular bridged systems, such as insulin, which has S-S bonds between different regions within the same chain [37]. Here, for the purpose of illustration, we apply our treatment to the porphyrin ring system shown in Fig. 4.7. This ring system, as well as other cyclic porphyrin arrays, can serve as models for photosynthetic light-harvesting antennae. Excellent reviews of recent progress in the synthesis of such materials are available [44–46].

The electronic spectrum is of strong interest with regard to the study of light-harvesting antennae because the former contains information about excitation energy hopping and the strength of exciton coupling between adjacent fragments. Time-dependent density functional theory (TD-DFT) has been applied to calculate excitation energies and oscillator strengths of zinc oligoporphyrins with various curved surface structures including ring, tube and ball shapes [47]. However, correlated ab initio methods of high quality cannot readily be applied to such large molecular assemblies even using highly advanced supercomputers. Special techniques, such as the elongation method, are required in order to provide a favorable computational cost for HF and post-HF level of theory. Clearly, NLO properties could also be of considerable value in this same context.

In the free-base porphyrin ring system shown in Fig. 4.7 the first porphyrin unit that is frozen in the elongation procedure must be re-activated when the active part approaches ring closure. The G-ELG method was developed to handle such a situation. As noted above it automatically recognizes the interaction and correctly redefines the frozen and active regions for the ELG-SCF step. In this case, G-ELG calculations were done with a value of 10^{-5} for the magnitude of the interaction between the active region and the previously frozen regions. The difference between the conventional (CNV) and G-ELG total energies turned out to be less than 10^{-7} a.u./atom. A CPU time versus number of units comparison for the energy calculation at each size is shown in the figure. The time saving of more than an order of magnitude for 16 units is quite satisfying. For G-ELG alone the total accumulated CPU time, taking into account all elongation steps, is also plotted.

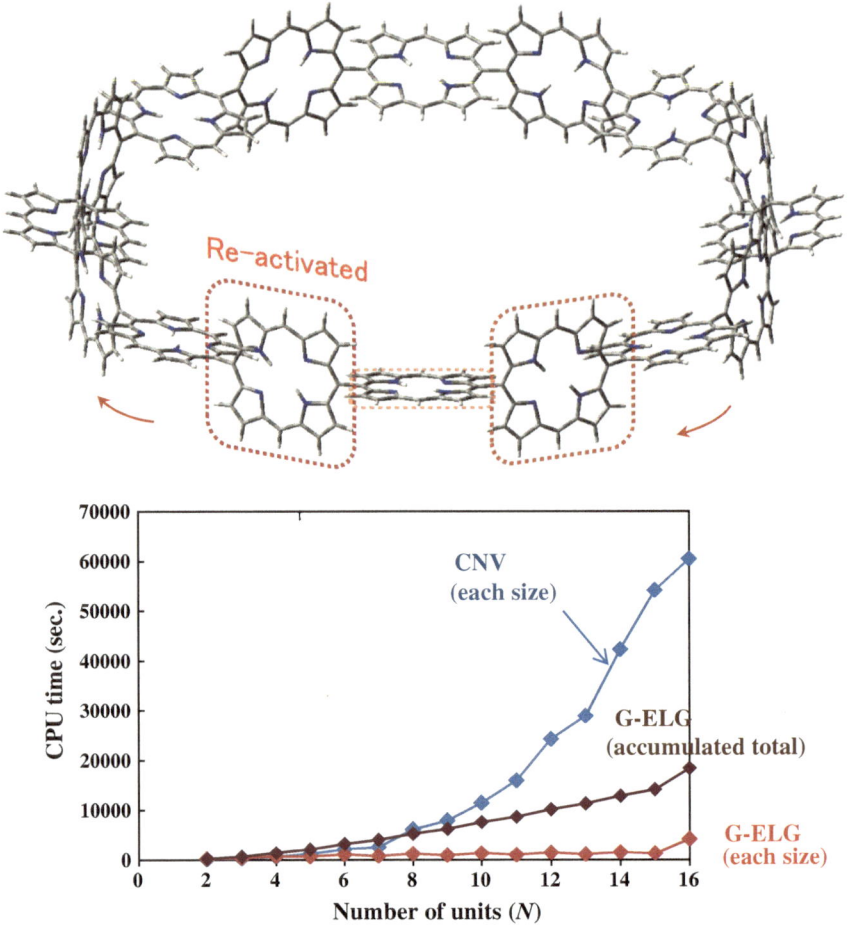

Fig. 4.7 Model of free base porphyrin ring for G-ELG method. CPU time comparison between CNV and G-ELG methods for each size, and also accumulated total CPU for G-ELG method

The G-ELG method is applicable to very large ring systems. In connection with NLO properties it would be interesting to monitor their values, along the chain direction, versus large N in order to compare with the infinite ring limit obtained by employing periodic boundary conditions as described in Chap. 3. It would also be of interest to examine light-harvesting properties calculated under similar conditions.

4.7.2 Por-(nT)-C_{60} Wire-Systems

Polymeric solar cells (PSCs) have been extensively studied as an alternative for producing renewable energy [48]. Photo-induced electron-transfer systems with highly efficient charge-separation (CS) and slow charge-recombination (CR) processes have

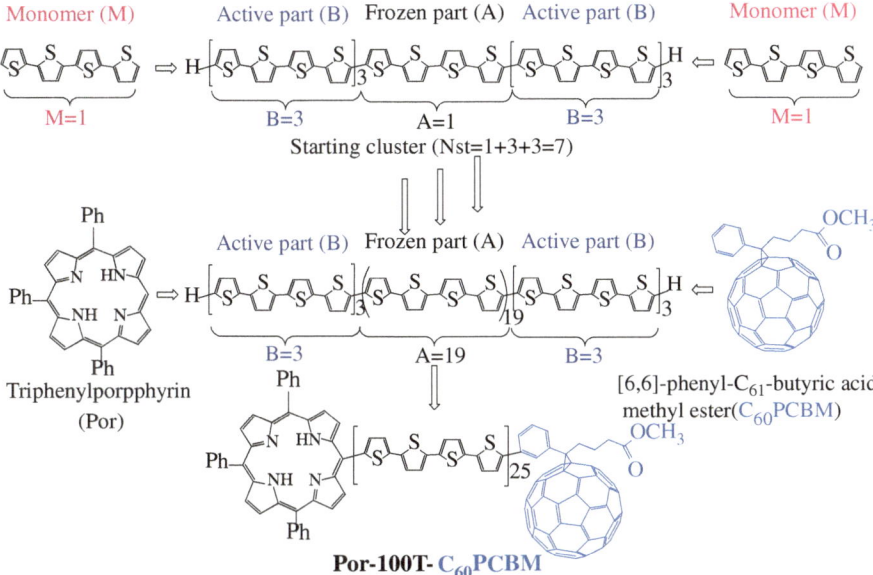

Fig. 4.8 Elongation process for Por-(nT)-C_{60} system

attracted considerable attention over past decade and are now being increasingly used in photovoltaic cells [49–51]. These systems have been synthesized and characterized, but there is yet no in-depth theoretical elucidation of the relevant charge transfer mechanisms. Efficient and reliable electronic structure calculations can play an important role in this regard and, ultimately, in the design of new functional materials. In that connection the G-ELG method has been applied to a model for the Triphenylporphyrin (Por)Oligothiophene (nT)Fullerene (C_{60}) photovoltaic cell. The scheme used to build this model by elongation is shown in Fig. 4.8 and the energy difference per atom with respect to a conventional calculation turned out to lie between 10^{-8} and 10^{-11} a.u. depending upon the size of the starting cluster. This demonstrates that the G-ELG method is effective for obtaining the electronic structures of such systems with high accuracy. The CPU time comparison for the elongation and conventional calculations is shown in Fig. 4.9. The last step corresponds to the attachment of large donor (Por) and acceptor (C_{60}) moieties and, therefore, the time increment is much greater than that required to add a single thiophene ring to the growing wire.

The photo-induced electron-transfer systems discussed in the preceding paragraph were synthesized in the presence of organic solvent. Otsubo and his colleagues [51] reported that the CS lifetime was related to the solvent polarity for the Por-nT-C_{60} system. In order to investigate the solvent effect one may use the popular polarizable continuum model (PCM) developed by Tomasi and co-workers [52], which can readily be combined with the ELG method. This was tested on the above Por-32T-C_{60} model using water and several organic solvents—benzonitrile (PhCN), o-dichlorobenzene (o-DCB) and toluene—with widely different dielectric

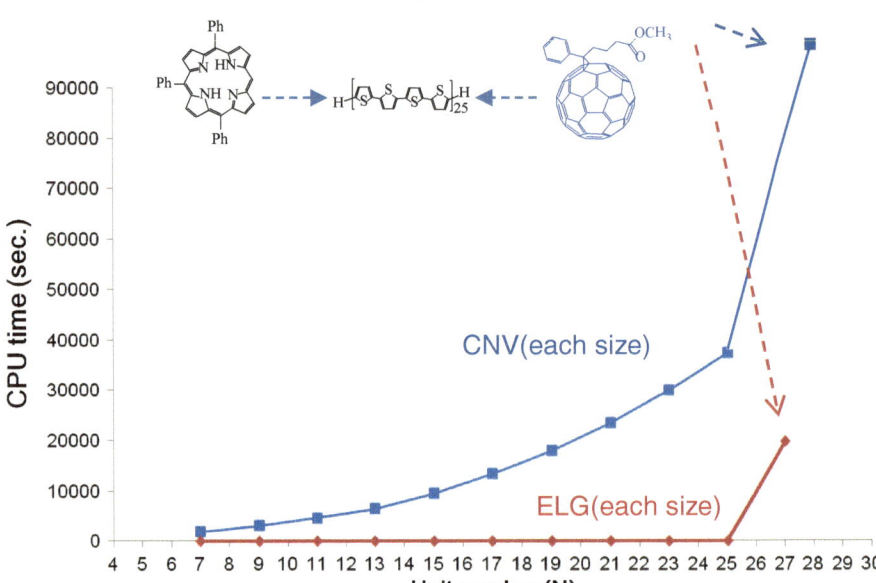

Fig. 4.9 CPU time comparison between elongation (ELG) and conventional (CNV) methods for the Por-(nT)-C_{60} system

Table 4.1 Effect of water and several organic solvents on the energy of Por-32T-C_{60} as calculated by the polarizable continuum model (PCM) using the conventional (CNV) and elongation (ELG) RHF/STO-3G methods

Solvent	Water	PhCN	o-DCB	Toluene
	($\varepsilon = 78.355$)	($\varepsilon = 25.592$)	($\varepsilon = 9.995$)	($\varepsilon = 2.374$)
CNV	−21471.56426	−21471.56426	−21471.56426	−21471.56426
CNV-PCM	−21471.60538	−21471.60422	−21471.60154	−21471.58782
CNV-ΔE_{solv}	−25.80175	−25.07274	−23.39317	−14.78362
ELG-PCM	−21471.56425	−21471.56425	−21471.56425	−21471.56425
CNV-PCM	−21471.60537	−21471.60420	−21471.60152	−21471.58781
ELG-ΔE_{solv}	−25.80288	−25.07386	−23.39210	−14.78432

In the elongation calculations the number of units in the B region is NBUNT=3 and the size of the starting cluster is Nst=12. The energy difference from gas phase to solvent, i.e. ΔE_{solv} is in kcal/mol whereas the total energies are in a.u

constants ε. Results for the effect of solvent on the energy at the RHF/STO-3G level are shown in Table 4.1. Once again, the differences from a conventional calculation (ELG-PCM vs. CNV-PCM) are small. The energy changes from gas phase to solvent for water are −25.80288 and −25.80175 kcal/mol for ELG-PCM and CNV-PCM, respectively. Similar differences were found for the three other solvents. Although the energy difference from gas phase to solvent increases with dielectric constant there is no simple relation between these two properties.

4.8 ELG-LMP2 Method

The elongation method is designed to describe the electronic structure of large molecular systems with good accuracy while keeping demand on computer resources at a reasonable level. Much effort has been devoted to the treatment of electron correlation for large systems at the ab initio wavefunction level by taking advantage of progress in computer technology and developments in methodology [53–55]. Basically, the various algorithms that have been presented exploit the near-sightedness of electronic interactions in one way or another. Two such approaches involve division of the entire system into subsystems and the use of localized orbitals, both of which are employed in the elongation method. Other methods based on the division into fragments include divide and conquer [56], fragment molecular orbitals [57], and the local space approximation [58]. At the MP2 level, perhaps the best known localized orbital methods are the Laplace-MP2 procedure of Häser and Almlöf [59], based on atomic orbitals, and the molecular orbital local MP2 (LMP2) procedure of Saebø and Pulay [60]. We follow the latter in using LMOs in the elongation method, but there are also significant differences that will be mentioned below.

The conventional MP2 method relies on the canonical molecular orbitals obtained from an HF calculation. In that basis the Fock matrix is diagonal, which leads to a simple expression for the second-order perturbation theory correction to the HF energy of a closed-shell system

$$E_{MP2} = \sum_{ij} \sum_{ab[ij]} [2T_{ab}^{ij} - T_{ba}^{ij}] K_{ab}^{ij} \qquad (4.24)$$

Here K_{ab}^{ij} denotes the two-electron integral

$$(ia|jb) = \int \varphi_i(1)^* \varphi_a(1) \frac{1}{r_{12}} \varphi_j^*(2) \varphi_b(2) d\mathbf{r_1} d\mathbf{r_2} \qquad (4.25)$$

with φ_i, φ_j being occupied CMOs and φ_a, φ_b virtual CMOs. The ij summation is over all the pairs of occupied orbitals for which correlation effects are considered, whereas the $ab[ij]$ summation is over all virtual orbital pairs that are employed to describe correlation effects for the (i, j) occupied pair. Later we will write just $[ij]$ instead of $ab[ij]$ and use the symbol $[i]$ to denote the virtual orbital domain for the ith occupied orbital. In Eq. (4.24), the quantity

$$T_{ab}^{ij} = \frac{(ai|bj)}{\varepsilon_i + \varepsilon_j - \varepsilon_a - \varepsilon_b}, \qquad (4.26)$$

in which $\varepsilon_i, \varepsilon_j, \varepsilon_a, \varepsilon_b$ are orbital energies, and $(ai|bj)$ is the amplitude of the (ij) to (ab) double excitation in the first-order perturbed wavefunction.

The evaluation of E_{MP2} as above leads to unfavorable scaling due to the fact that the canonical orbitals are strongly delocalized. Indeed, the transformation of the two-electron integrals from AOs to CMOs scales as $O(N^4) - O(N^5)$. On the other hand, it is known from numerical experiments that correlation in insulators is a very local effect decaying with inter-electron distance as r^{-6}. Thus, we conclude that high-order scaling is not an inherent feature of the MP2 approach, but rather results from a computationally unsuitable orbital basis set choice. One way of resolving this computational bottleneck is through the LMP2 methodology introduced by Saebø and Pulay [60]. They exploit the locality of correlation effects by making explicit use of LMOs in constructing the active space for treatment of the electron pair correlations. Their formalism utilizes the Hylleraas functional to provide a variational upper bound for E_{MP2}. The resulting formula may be expressed in exactly the same form as Eq. (4.15) except that the orbitals φ_i, φ_j are LMOs, while a set of redundant projected AOs (to remove occupied orbitals) is employed for description of the virtual orbitals φ_a and φ_b. The amplitudes are, then, determined from:

$$\underline{\underline{0}} = \underline{\underline{K}}^{ij} + \underline{\underline{F}} \, \underline{\underline{T}}^{ij} \, \underline{\underline{S}} + \underline{\underline{S}} \, \underline{\underline{T}}^{ij} \, \underline{\underline{F}} - \underline{\underline{S}} \sum_k [\underline{\underline{F}}^{ik} \, \underline{\underline{T}}^{kj} + \underline{\underline{T}}^{ik} \, \underline{\underline{F}}^{kj}] \underline{\underline{S}} \qquad (4.27)$$

in which $\underline{\underline{K}}^{ij}$ and $\underline{\underline{T}}^{ij}$ are square matrices indexed by the virtual orbital pairs belonging to $[i\overline{j}]$ and $\underline{\underline{F}}$ and $\underline{\underline{S}}$ are the corresponding virtual-virtual Fock and overlap AO matrices, respectively. In the ELG-MP2 treatment the virtual AO basis is replaced by a virtual LMO basis (see later) and the above equation simplifies since $\underline{\underline{S}}$ becomes the identity. In either case Eq. (4.27) is a set of matrix equations for the amplitude matrices $\underline{\underline{T}}^{ij}$ that are coupled through the non-diagonal Fock matrices $\underline{\underline{F}}^{ik}$. As a result, the $\underline{\underline{T}}^{ij}$ have to be determined iteratively, which is done by means of the Gauss-Seidel method. In ELG-LMP2 calculations we have observed that typically, similar to LMP2, not more than 10 iterations are needed to reach convergence of order 10^{-7} in the energy.

If all occupied pairs are taken into account in the above procedure, and the full virtual space is used for each occupied pair, then the ELG-LMP2 (or LMP2) method will give the same result as a standard MP2 calculation, assuming the preliminary ELG-HF treatment is 'exact'. However, that would be a very time-consuming process. Thus, in ELG-LMP2 we assume, in addition, that correlation is significant for an (i, j) pair only if the i and j orbitals are 'spatially near' (see later) to each other. We also assume that virtual orbital a has to be included in $[i]$, only if a and i are spatially near to one another. The virtual pair domains $[ij]$ are, then, constructed as symmetrized direct products of $[i]$ and $[j]$. With these assumptions the number of significant (i, j) pairs grows linearly with system size for large systems, while the average size of $[ij]$ remains constant. As a result, the number of amplitudes T_{ab}^{ij}, as well as the number of integrals in the LMO basis and the number of equations to be solved, scales as $O(N)$.

The possible computational bottlenecks that remain to be discussed concern the number of two-electron AO integrals that must be calculated and the scaling for the transformation of these integrals to the RLMO basis. ELG-LMP2 includes a cut-off technique that reduces the number of AO non-zero integrals to $O(N)$ [21, 22]. There remains the integrals transformation, which is simplified by taking advantage of the fact that, in both conventional LMP2 and ELG-LMP2, the correlating electron pairs are limited to those involving LMOs that are 'spatially near' to one another. However, the criterion used is much different in the two methods. In ELG-LMP2 it is based on the RLMOs obtained in the ELG-SCF calculation. The two-electron matrix elements for a given (i, j) pair of occupied orbitals are included in the calculations only if i and j are localized in the same or neighbouring regions. Similarly, we assume that the virtual orbital $a \in [i]$ only if i and a are localized in the same or neighbouring regions. It is found that very good accuracy can be obtained if only orbitals of the nearest and the second-nearest neighbours are included in the calculation. Of course, the treatment can be extended to include more neighbors should higher accuracy be needed.

As far as the two-electron transformation from AOs to RLMOs is concerned, we divide this process (as usual) into four nested steps,

$$(ia|jb) = \sum_{\alpha} C_{\alpha i} \sum_{\beta} C_{\beta a} \sum_{\gamma} C_{\gamma j} \sum_{\delta} C_{\delta b} (\alpha\beta|\gamma\delta). \qquad (4.28)$$

Schwartz's inequality gives that

$$|(ia|jb)| \leq [(ia|ia)(jb|jb)]^{1/2}. \qquad (4.29)$$

Thus, when any of the terms on the rhs is below a pre-set threshold, the left-hand is not calculated. In addition, permutation symmetry is used to reduce the number of integrals to be considered. Moreover, if $(\alpha\beta|\gamma\delta)$ all belong to the frozen region, $(i, j$ denote occupied and a, b virtual RLMOs that were localized to the active region), the contribution of i, j, a, b to $(\alpha\beta|\gamma\delta)$ is negligible and the matrix elements $(ia|jb)$ is close to 0. Then, this specific transformation can be skipped. As a result of these procedures one obtains a transformation that scales nearly linearly for both CPU time and memory requirements.

ELG-LMP2 has now been linked to the implementation of the elongation method in the GAMESS package [15]. Test calculations can be found in Ref. [39].

4.9 ELG-LCIS Method

The elongation local configuration interaction singles (ELG-LCIS) method for calculating excited electronic states, like ELG-LMP2, is based on taking advantage of 'spatial nearness'. In this case the RLMOs of the entire system, obtained in the initial

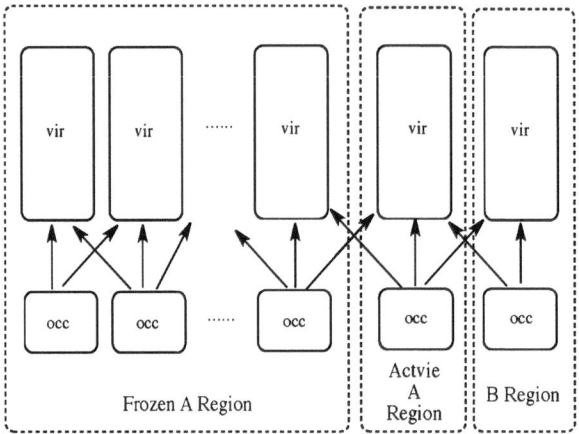

Fig. 4.10   The nearest neighbor excitations included in the ELG-LCIS method. In this method, the RLMOs are divided into occ-vir regions. The arrows show all the allowed excitations

ELG-SCF calculation, are divided into regions, each with local occupied and virtual orbitals. In contrast with the traditional CIS method, in which excitations from all occupied orbitals to all virtual orbitals are considered, in ELG-LCIS we consider only excitations within a single region or from occupied orbitals within that region to virtual orbitals in the nearest neighbor region(s) as shown in Fig. 4.10.

We can see that the RLMOs have been divided into many parts. In doing so the total number of configurations included in the ELG-LCIS expansion, including excitations from all occupied orbitals, is dramatically reduced relative to a conventional CIS calculation. Consequently, ELG-LCIS can achieve a large speedup over conventional CIS provided that the code is optimized and utilizes high performance computing (HPC).

Table 4.2 presents CIS and ELG-LCIS results obtained with the 6-31G basis set for the lowest-lying set of electronic transitions in a 12 unit non-bonded benzene system and an 18 unit bonded polyene at the equilibrium geometry. As one can see

Table 4.2   Lowest lying electronic excitation energies (a.u.) of face-to-face stacked (benzene)$_{12}$ (non-bonding system) and polyene with 18 C=C units (bonded system) from ELG-LCIS and conventional CIS calculations

	Configurations	State 1	State 2	State 3	State 4
(benzene)$_{12}$					
CNV-CIS	136080	0.16054507	0.16087162	0.16372588	0.16447498
ELG-CIS	52704	0.16244784	0.16285274	0.16551854	0.1663486
Polyene					
CNV-CIS	76035	0.12587582	0.12605687	0.19014543	0.19025067
ELG-CIS	35000	0.12587580	0.12605845	0.19025068	0.19046077

from the table the number of excitation configurations is reduced in comparison with conventional CIS by almost a factor of 3 for the benzene system and by a factor of 2 for the polyene. The accuracy is within 1 % for the excitation energy.

It should be mentioned that ELG-LCIS requires high quality localization for the RLMOs. Moreover, the treatment needs to be extended to include double excitations based on the ideas used in ELG-LMP2.

References

1. Löwdin, P.-O.: Quantum theory of cohesive properties of solids. Adv. Phys. **5**, 1–171 (1956)
2. Re, G.D., Ladik, L., Biczó, G.: Self-consistent-field tight-binding treatment of polymers. I. Infinite three-dimensional case. Phys. Rev. **155**, 997–1003 (1967)
3. André, J.-M., Gouverneur, L., Leroy, G.: L'Etude théorique des systémes périodiques. I. La méthode LCAO-HCO. Int. J. Quant. Chem. **1**, 427–450 (1967)
4. André, J.-M., Gouverneur, L., Leroy, G.: L'Étude théorique des systémes périodiques. II. La méthode LCAO-SCF-CO. Int. J. Quant. Chem. **1**, 451–461 (1967)
5. Pisani, C.: Ab-initio approaches to the quantum-mechanical treatment of periodic systems. Lect. Notes Chem. **67**, 47–75 (1996)
6. Dovesi, R., Civalleri, B., Roetti, C., Saunders, V.R., Orlando, R.: Ab initio quantum simulation in solid state chemistry. Rev. Comput. Chem. **21**, 1–125 (2005)
7. Imamura, A., Aoki, Y., Maekawa, K.: A theoretical synthesis of polymers by using uniform localization of molecular orbitals: proposal of an elongation method. J. Chem. Phys. **95**, 5419–5431 (1991)
8. Maekawa, K., Imamura, A.: Electronic structures around the local defects in all-trans-polyacetylene: an analysis by the cluster-series model. Int. J. Quant. Chem. **47**, 449–467 (1993)
9. Kurihara, Y., Aoki, Y., Imamura, A.: Calculations of the excitation energies of all-trans and 11, 12s-dicis retinals using localized molecular orbitals obtained by the elongation method. J. Chem. Phys. **107**, 3569–3575 (1997)
10. Aoki, Y., Suhai, S., Imamura, A.: An efficient cluster elongation method in density functional theory and its application to poly-hydrogen-bonding molecules. J. Chem. Phys. **101**, 10808–10823 (1994)
11. Aoki, Y., Suhai, S., Imamura, A.: A density functional elongation method for the theoretical synthesis of aperiodic polymers. Int. J. Quant. Chem. **52**, 267–280 (1994)
12. Mitani, M., Aoki, Y., Imamura, A.: Electronic structures of large, extended, nonperiodic systems by using the elongation method: model calculations for the cluster series of a polymer and the molecular stacking on a surface. Int. J. Quant. Chem. **54**, 167–196 (1995)
13. Mitani, M., Imamura, A.: A general quantum chemical approach to study the locally perturbed periodic systems: a new development of the ab initio crystal elongation method. J. Chem. Phys. **103**, 663–675 (1995)
14. Gu, F.L., Aoki, Y., Korchowiec, J., Imamura, A., Kirtman, B.: A new localization scheme for the elongation method. J. Chem. Phys. **121**, 10385–10391 (2004)
15. Schmidt, M.W., Baldridge, K.K., Boatz, J.A., Elbert, S.T., Gordon, M.S., Jensen, J.H., Koseki, S., Matsunaga, N., Nguyen, K.A., Su, S., Windus, T.L., Dupuis, M., Montgomer, J.A.J.: General atomic and molecular electronic structure system. J. Comput. Chem. **14**, 1347–1363 (1993)
16. Steinborn, E.O., Ruedenberg, K.: Rotation and translation of regular and irregular solid spherical harmonics. Adv. Quant. Chem. **7**, 1–81 (1973)
17. Greengard, L.F.: The Rapid Evaluation of Potential Fields in Particle Systems. MIT, Cambridge (1987)
18. Choi, C.H., Ivanic, J., Gordon, M.S., Ruedenberg, K.: Rapid and stable determination of rotation matrices between spherical harmonics by direct recursion. J. Chem. Phys. **111**, 8825–8831 (1999)

19. Choi, C.H., Ruedenberg, K., Gordon, M.S.: New parallel optimal-parameter fast multipole method (OPFMM). J. Comput. Chem. **22**, 1484–1501 (2001)
20. Choi, C.H.: Direct determination of multipole moments of Cartesian Gaussian functions in spherical polar coordinates. J. Chem. Phys. **120**, 3535–3543 (2004)
21. Korchowiec, J., Lewandowski, J., Makowski, M., Gu, F.L., Aoki, Y.: Elongation cutoff technique armed with quantum fast multipole method for linear scaling. J. Comput. Chem. **30**, 2515–2525 (2009)
22. Korchowiec, J., Silva, P.D., Makowski, M., Gu, F.L., Aoki, Y.: Elongation cutoff technique at Kohn-Sham level of theory. Int. J. Quant. Chem. **110**, 2130–2139 (2010)
23. Pomogaeva, A., Kirtman, B., Gu, F.L., Aoki, Y.: Band structure built from oligomer calculations. J. Chem. Phys. **128**, 074109 (2008)
24. Pomogaeva, A., Springborg, M., Kirtman, B., Gu, F.L., Aoki, Y.: Band structures built by the elongation method. J. Chem. Phys. **130**, 194106 (2009)
25. Pomogaeva, A., Gu, F.L., Imamura, A., Aoki, Y.: Electronic structures and nonlinear optical properties of supramolecular associations of benzo-2,1,3-chalcogendiazoles by the elongation method. Theor. Chem. Acc. **125**, 453–460 (2010)
26. Orimoto, Y., Gu, F.L., Imamura, A., Aoki, Y.: Efficient and accurate calculations on the electronic structure of B-type poly(dG)·poly(dC) DNA by elongation method: first step toward the understanding of the biological properties of aperiodic DNA. J. Chem. Phys. **126**, 215104 (2007)
27. Ohnishi, S., Gu, F.L., Naka, K., Imamura, A., Kirtman, B., Aoki, Y.: Calculation of static (hyper)polarizabilities for π-conjugated donor/acceptor molecules and block copolymers by the elongation finite-field method. J. Phys. Chem. A **108**, 8478–8484 (2004)
28. Gu, F.L., Champagne, B., Aoki, Y.: Evaluation of nonlinear susceptibilities of 3-methyl-4-nitropyridine 1-oxide crystal: an application of the elongation method to nonlinear optical properties. Lect. Ser. Comput. Comput. Sci. Eng. **1**, 779–782 (2004)
29. Gu, F.L., Guillaume, M., Botek, E., Champagne, B., Castet, F., Ducasse, L., Aoki, Y.: Elongation method and supermolecule approach for the calculation of nonlinear susceptibilities. Application to the 3-methyl-4-nitropyridine 1-oxide and 2-Methyl-4-nitroaniline crystals. J. Comput. Meth. Sci. Eng. **6**, 171–188 (2006)
30. Ohnishi, S., Gu, F.L., Naka, K., Aoki, Y.: Parallelization efficiency of the elongation method and its application to NLO design for urea crystal. Comput. Lett. **3**, 231–241 (2007)
31. Ohnishi, S., Orimoto, Y., Gu, F.L., Aoki, Y.: Nonlinear optical properties of polydiacetylene with donor-acceptor substitution block. J. Chem. Phys. **127**, 084702 (2007)
32. Chen, W., Yu, G.-T., Gu, F.L., Aoki, Y.: Investigation on the electronic structures and nonlinear optical properties of pristine boron nitride and boron nitride/carbon heterostructured single-wall nanotubes by the elongation method. J. Phys. Chem. C **113**, 8447–8454 (2009)
33. Chen, W., Yu, G.-T., Gu, F.L., Aoki, Y.: Investigation on nonlinear optical properties of ladder-structure polydiacetylenes derivatives by using the elongation finite-field method. Chem. Phys. Lett. **474**, 175–179 (2009)
34. Yan, L.K., Pomogaeva, A., Gu, F.L., Aoki, Y.: Theoretical study on nonlinear optical properties of metalloporphyrin using elongation method. Theor. Chem. Acc. **125**(3–6), 511–520 (2010)
35. Aoki, Y., Gu, F.L.: Generalized elongation method: from one-dimension to three-dimension. AIP Conf. Proc. **1504**, 647–650 (2012)
36. Aoki, Y., Gu, F.L.: An elongation method for large systems toward bio-systems. Phys. Chem. Chem. Phys. **14**, 7640–7668 (2012)
37. Liu, K., Peng, L., Gu, F.L., Aoki, Y.: Three dimensional elongation method for large molecular calculations. Chem. Phys. Lett. **560**, 66–70 (2013)
38. Liu, K., Yan, Y.-A., Gu, F.L., Aoki, Y.: A modified localization scheme for the three-dimensional elongation method applied to large systems. Chem. Phys. Lett. **565**, 143–147 (2013)
39. Makowski, M., Korchowiec, J., Gu, F.L., Aoki, Y.: Describing electron correlation effects in the framework of the elongation method-elongation-MP2: formalism, implementation and efficiency. J. Comput. Chem. **31**, 1733–1740 (2010)

40. Makowski, M., Gu, F.L., Aoki, Y.: Elongation-CIS method: describing excited states of large molecular systems in regionally localized molecular orbital basis. J. Comput. Meth. Sci. Eng. **10**, 473–481 (2010)

41. Löwdin, P.-O.: On the non-orthogonality problem connected with the use of atomic wave functions in the theory of molecules and crystals. J. Chem. Phys. **18**, 365–375 (1950)

42. Reed, A.E., Weinhold, F.: Natural localized molecular orbitals. J. Chem. Phys. **83**, 1736–1740 (1985)

43. Aoki, Y., Loboda, O., Liu, K., Makowski, M.A., Gu, F.L.: Highly accurate O(N) method for delocalized systems. Theor. Chem. Acc. **130**(4–6), 595–608 (2011)

44. Aratani, N., Kim, D., Osuka, A.: Discrete cyclic porphyrin arrays as artificial light-harvesting antenna. Acc. Chem. Res. **42**, 1922–1934 (2009)

45. Tsuda, A., Nakamura, T., Sakamoto, S., Yamaguchi, K., Osuka, A.: A self-assembled porphyrin box from mesomeso-linked bis[5-p-pyridyl-15-(3,5-di-octyloxyphenyl)porphyrinato zinc(II)]. Angew. Chem. Int. Ed. **41**, 2817–2821 (2002)

46. Yoon, M.-C., Cho, S., Kim, P., Hori, T., Aratani, N., Osuka, A., Kim, D.: Structural dependence on excitation energy migration processes in artificial light harvesting cyclic zinc(II) porphyrin arrays. J. Phys. Chem. B **113**, 15074–15082 (2009)

47. Yamaguchi, Y.: Theoretical prediction of electronic structures of fully π-conjugated zinc oligoporphyrins with curved surface structures. J. Chem. Phys. **120**, 7963–7970 (2004)

48. Cheng, Y.-J., Yang, S.-H., Hsu, C.-S.: Synthesis of conjugated polymers for organic solar cell applications. Chem. Rev. **109**, 5868–5923 (2009)

49. Ikemoto, J., Takimiya, K., Aso, Y., Otsubo, T., Fujitsuka, M., Ito, O.: Porphyrino-ligothiophenefullerene triads as an efficient intramolecular electron-transfer system. Org. Lett. **4**, 309–311 (2002)

50. Nakamura, T., Ikemoto, J., Fujitsuka, M., Araki, Y., Ito, O., Takimiya, K., Aso, Y., Otsubo, T.: Control of photoinduced energy- and electron-transfer steps in zinc porphyrinoligothiophene-fullerene linked triads with solvent polarity. J. Phys. Chem. B **109**, 14365–14374 (2005)

51. Nakamura, T., Ikemoto, J., Fujitsuka, M., Araki, Y., Ito, O., Takimiya, K., Aso, Y., Otsubo, T.: Control of photoinduced energy- and electron-transfer steps in zinc photoinduced electron transfer in porphyrin-oligothiophene-fullerene linked triads with solvent polarity. J. Phys. Chem. B **108**, 10700–10710 (2004)

52. Miertuš, S., Scrocco, E., Tomasi, J.: Electrostatic interaction of a solute with a continuum. A direct utilization of ab initio molecular potentials for the prevision of solvent effects. Chem. Phys. **55**, 117–129 (1981)

53. Shavitt, I., Bartlett, R.J.: Many-Body Methods in Chemistry and Physics. MBPT and Coupled-Cluster Theory. Cambridge University Press, Cambridge (2009)

54. Werner, H.-J., Manby, F.R., Knowles, P.J.: Fast linear scaling second-order Møller-Plesset perturbation theory (MP2) using local and density fitting approximations. J. Chem. Phys. **118**, 8149–8160 (2003)

55. Scuseria, G.E.: Linear scaling density functional calculations with gaussian orbitals. J. Phys. Chem. A **103**, 4782–4790 (1999)

56. Yang, W.: Direct calculation of electron density in density-functional theory. Phys. Rev. Lett. **66**, 1438–1441 (1991)

57. Kitaura, K., Ikeo, E., Asada, T., Nakano, T., Uebayasi, M.: Fragment molecular orbital method: an approximate computational method for large molecules. Chem. Phys. Lett. **313**, 701–706 (1999)

58. Kirtman, B., de Melo, C.: Density matrix treatment of localized electronic interactions in molecules and solids. J. Chem. Phys. **75**, 4592–4602 (1981)

59. Häser, M., Almlöf, J.: Laplace transform techniques in Møller-Plesset perturbation theory. J. Chem. Phys. **96**, 489–494 (1992)

60. Saebø, S., Pulay, P.: Local treatment of electron correlation. Annu. Rev. Phys. Chem. **44**, 213–236 (1993)

Chapter 5
Applications of the Elongation Method to NLO Properties

Abstract In this chapter we present our methodologies for calculating L&NLO properties of large systems using the ELG approach. In addition to the issue of linear scaling, the accuracy of the calculated field-free, finite field and NLO properties is established through comparison with conventional finite-system and infinite-periodic-system calculations. Results are presented for quasilinear systems including conjugated polymers, nanotubes, nanowires, etc. A second set of calculations is presented to demonstrate the applicability of the ELG to large systems that are not quasilinear. For such systems there may be significant interactions between fragments that are not directly connected in the build-up process. We show that these interactions are properly taken into account within the framework of the ELG method.

5.1 Introduction

In this chapter, the ELG-SCF procedures for treating a large system in the presence of an external electric field are given. Both the ELG-FF (Sect. 5.2) and the ELG-CPHF (Sect. 5.3) methods are developed as extensions of the field-free procedure. Illustrative static field calculations are presented in Sect. 5.4 for each of these methods. It is seen that the G-ELG procedure works well for NLO properties even for strongly delocalized systems.

5.2 ELG-FF Method

Again we concentrate on the effects of a spatially uniform (homogeneous), external, electric field $\mathbf{E}$. The interaction of the electrons with the field will be described by the scalar potential discussed in Chap. 3. In that case, the field-free Hamiltonian is augmented by a term $\hat{H}' = -e\mathbf{E} \cdot \mathbf{r}$ for each electron. The contribution from the field acting on the nuclei may be treated simply as an additive constant contribution to the total energy and, hence, will be omitted in the following.

© The Author(s) 2015   67
F.L. Gu et al., *Calculations on nonlinear optical properties for large systems*,
SpringerBriefs in Electrical and Magnetic Properties of Atoms,
Molecules, and Clusters, DOI 10.1007/978-3-319-11068-4_5

There are two different approaches for carrying out a conventional Hartree-Fock (or Kohn-Sham) calculation aimed at studying the interaction with the electrostatic field, and also within the elongation method this is the case. One of these approaches, discussed in Chap. 2, is the numerical finite field (FF) procedure, which may be directly incorporated within the framework of the ELG method. It is the most straight-forward approach, but is limited to static fields and the results may depend critically on the chosen field strengths. In the FF method one proceeds in exactly the same manner as in the absence of a field except for the additional one-electron term in the Hamiltonian and the fact that all quantities depend upon the magnitude and direction of the field as parameters. Thus, in the elongation finite field method (ELG-FF) localization is carried out in exactly the same manner as previously described and the Hartree-Fock (Kohn-Sham) SCF equations (4.20), (4.21), and (4.22) take exactly the same form except that $\underline{\underline{H}}^{\text{core}}$ is replaced by $\underline{\underline{H}}^{\text{core}} + \underline{\underline{H}}'$. It is subsequently assumed that the resulting energy in Eq. (4.22) may be expanded as a power series in the field

$$
E_{\text{tot}}(\mathbf{E}) = E_{\text{tot}}(0) - \sum_i \mu_i E_i - \frac{1}{2!} \sum_{ij} \alpha_{ij} E_i E_j - \frac{1}{3!} \sum_{ijk} \beta_{ijk} E_i E_j E_k
$$
$$
- \frac{1}{4!} \sum_{ijkl} \gamma_{ijkl} E_i E_j E_k E_l + \cdots , \tag{5.1}
$$

where the subscripts indicate Cartesian components, μ_i is the ith component of the dipole moment, and the (hyper)polarizabilities are the derivatives:

$$
\alpha_{ij} = \left(\frac{\partial^2 E_{\text{tot}}(\mathbf{E})}{\partial E_i \partial E_j} \right)_{\mathbf{E}=0}
$$
$$
\beta_{ijk} = \left(\frac{\partial^3 E_{\text{tot}}(\mathbf{E})}{\partial E_i \partial E_j \partial E_k} \right)_{\mathbf{E}=0}
$$
$$
\gamma_{ijkl} = \left(\frac{\partial^4 E_{\text{tot}}(\mathbf{E})}{\partial E_i \partial E_j \partial E_k \partial E_l} \right)_{\mathbf{E}=0} \tag{5.2}
$$

Based on total energies obtained for fields of strengths $\pm E$, $\pm 2E$ in the same direction one may calculate the diagonal components of the static (hyper)polarizabilities in that direction as:

$$
\alpha = \frac{1}{E^2} \{ \frac{5}{2} E_{\text{tot}}(0) - \frac{4}{3} [E_{\text{tot}}(+E) + E_{\text{tot}}(-E)] + \frac{1}{12} [E_{\text{tot}}(+2E) + E_{\text{tot}}(-2E)] \}
$$
$$
\beta = \frac{1}{E^3} \{ E_{\text{tot}}(+E) - E_{\text{tot}}(-E) - \frac{1}{2} [E_{\text{tot}}(+2E) - E_{\text{tot}}(-2E)] \}
$$
$$
\gamma = \frac{1}{E^4} \{ -6E_{\text{tot}}(0) + 4[E_{\text{tot}}(+E) + E_{\text{tot}}(-E)] - [E_{\text{tot}}(+2E) + E_{\text{tot}}(-2E)] \}. \tag{5.3}
$$

The entire set of tensor components may be obtained simply by considering different field directions.

5.3 ELG-CPHF/CPKS Method

In order to obtain dynamic (i.e., frequency-dependent) NLO properties we apply time-dependent perturbation theory at the coupled perturbed Hartree-Fock (CPHF) or coupled perturbed Kohn-Sham (CPKS) level. The conventional CPHF/CPKS procedure has been presented in detail in Sect. 2.4. Here we emphasize the adaptation of the elongation method to dovetail with conventional perturbation theory. We start with the localization procedure of Sect. 4.2.

The electric field **E** enters the localization procedure through the density matrix in the OAO representation $\underline{\underline{D}}^{OAO}$, which becomes field-dependent. This matrix determines the transformation from OAOs to ROs through the matrix $\underline{\underline{T}}$ in Eq. (4.9). In the field-free case, $\underline{\underline{T}}^A$ is the matrix of eigenvectors of $\underline{\underline{D}}^{OAO}(A)$, with eigenvalues $\underline{\underline{\lambda}}^A$

$$\underline{\underline{D}}^{OAO}(A)\,\underline{\underline{T}}^A = \underline{\underline{T}}^A\,\underline{\underline{\lambda}}^A \tag{5.4}$$

and $\underline{\underline{T}}^B$ is defined analogously. The logical way to proceed is to include the effect of the field perturbatively. However, the diagonalization of $\underline{\underline{D}}^{OAO(A)}$ and $\underline{\underline{D}}^{OAO(B)}$ cannot be carried out by ordinary perturbation theory because the field-free eigenvalues are quasi-degenerate. In fact, they fall into two groups with values that are either close to 2.0 for approximately occupied orbitals and close to 0.0 for approximately unoccupied orbitals. This means we can apply ordinary perturbation theory to eliminate the off-diagonal elements that couple the approximately occupied with the approximately unoccupied blocks of $\underline{\underline{D}}^{OAO}$ but not to remove the coupling within the diagonal blocks [1]. Accordingly, in the presence of a field we require that $\underline{\lambda}$ be block diagonal rather than fully diagonal. Although the field thereby mixes the localized orbitals within each region it does not affect the overall regional localization. The mathematical analysis is very similar to what occurs in the non-canonical treatment of the perturbed Fock equation [2] described in Sect. 2.4 except, in that case, it is the matrix of undetermined Lagrange multipliers that is taken to be block-diagonal. Thus, an exactly analogous procedure is followed below wherein the $\underline{\lambda}$ matrices and the unitary $\underline{\underline{T}}$ matrices are expanded as Taylor series in the field in such a manner that quasi-degeneracies are avoided.

For simplicity, we first derive the equations for the case of static fields and, at the end, generalize to the case of dynamic fields. Thus,

$$\underline{\underline{D}}^{OAO}(A) = \underline{\underline{D}}^{(0)OAO}(A) + E\underline{\underline{D}}^{(1)OAO}(A) + E^2\underline{\underline{D}}^{(2)OAO}(A) + \cdots, \tag{5.5}$$

with

$$\underline{\underline{T}}^A = \underline{\underline{T}}^{(0)A} + E\underline{\underline{T}}^{(1)A} + E^2\underline{\underline{T}}^{(2)A} + \cdots, \tag{5.6}$$

and

$$\underline{\underline{\lambda}}^A = \underline{\underline{\lambda}}^{(0)A} + E\underline{\underline{\lambda}}^{(1)A} + E^2\underline{\underline{\lambda}}^{(2)A} + \cdots \tag{5.7}$$

Inserting these expansions into Eq. (5.4) leads to the first order expression

$$\underline{\underline{D}}^{(0)OAO}(A)\,\underline{\underline{T}}^{(1)A} + \underline{\underline{D}}^{(1)OAO}(A)\,\underline{\underline{T}}^{(0)A} = \underline{\underline{T}}^{(0)A}\,\underline{\underline{\lambda}}^{(1)A} + \underline{\underline{T}}^{(1)A}\,\underline{\underline{\lambda}}^{(0)A} \tag{5.8}$$

In Eq. (5.5) $\underline{\underline{D}}^{(1)OAO}(A)$, $\underline{\underline{D}}^{(2)OAO}(A)$, $\ldots$ are obtained as described in Sect. 2.4. It is convenient to write $\underline{\underline{T}}^{(1)A}$ as

$$\underline{\underline{T}}^{(1)A} = \underline{\underline{T}}^{(0)A}\,\underline{\underline{\eta}}^{(1)A} \tag{5.9}$$

Then, multiplying Eq. (5.8) on the left by $\underline{\underline{T}}^{(0)A^\dagger}$ and using the fact that $\underline{\underline{T}}^{(0)A}$ is unitary, results in

$$\underline{\underline{\lambda}}^{(0)A}\,\underline{\underline{\eta}}^{(1)A} + \underline{\underline{g}}^{(1)A} + \underline{\underline{\eta}}^{(1)A^\dagger}\,\underline{\underline{\lambda}}^{(0)A} = \underline{\underline{\lambda}}^{(1)A} \tag{5.10}$$

where $\underline{\underline{g}}^{(1)A} = \underline{\underline{T}}^{(0)A^\dagger}\,\underline{\underline{D}}^{(1)OAO}(A)\underline{\underline{T}}^{(0)A}$. Since $\underline{\underline{\lambda}}^{(1)}$ is defined to be block-diagonal, the matrix elements of $\underline{\underline{\eta}}^{(1)A}$ that couple the off-diagonal blocks are given by

$$\eta_{ia}^{(1)A} = \frac{g_{ia}^{(1)A}}{\lambda_a^{(0)A} - \lambda_i^{(0)A}} \tag{5.11}$$

with i being an approximately doubly-occupied orbital and a being an approximately empty orbital. In order to preserve orthonormality we require that $\underline{\underline{T}}^A$ remains unitary for all fields, i.e., $\underline{\underline{T}}^{(0)A}\,\underline{\underline{T}}^{(1)A^\dagger} + \underline{\underline{T}}^{(1)A}\,\underline{\underline{T}}^{(0)A^\dagger} = \underline{\underline{0}}$. Therefore, $\underline{\underline{\eta}}^{(1)A}$ in Eq. (5.9) must satisfy

$$\underline{\underline{\eta}}^{(1)A} + \underline{\underline{\eta}}^{(1)A^\dagger} = \underline{\underline{0}}. \tag{5.12}$$

Equation (5.11) is consistent with that choice as far as the matrix elements of the off-diagonal blocks of $\underline{\underline{\eta}}^{(1)A}$ are concerned. For the diagonal blocks we make the choice $\underline{\underline{\eta}}^{(1)A} = \underline{\underline{0}} = \underline{\underline{\eta}}^{(1)A^\dagger}$ which, to be consistent with Eq. (5.10), requires that the diagonal blocks of $\underline{\lambda}$ are given by $[\lambda^{(1)A)}]_{ij} = g_{ij}$ and $[\lambda^{(1)A)}]_{ab} = g_{ab}$. That completes the derivations of the expressions for $\underline{\underline{T}}^{(1)A}$ and $\underline{\lambda}^{(1)A}$, i.e., the first-order corrections in the external field strength. The $\underline{\underline{T}}^{(1)B}$ and $\underline{\lambda}^{(1)B}$ matrices can be obtained similarly, and the total $\underline{\underline{T}}^{(1)}$ is the direct sum of $\underline{\underline{T}}^{(1)A}$ and $\underline{\underline{T}}^{(1)B}$,

$$\underline{\underline{T}}^{(1)} = \underline{\underline{T}}^{(1)A} \oplus \underline{\underline{T}}^{(1)B}. \tag{5.13}$$

The fact that $\underline{\underline{\lambda}}^{(1)}$ is only block diagonal affects the higher-order solutions but, as noted above, does not affect the regional localization.

Next we turn to the second step in the localization, which is designed to preserve the key feature of the zero-field RLMOs: that they are fully occupied/unoccupied and maintain the localization of the ROs as much as possible. In order to achieve this the field-dependent coupling between the occupied and unoccupied blocks in the first-order RO density matrix,

$$\underline{\underline{D}}^{(1)RO} = \underline{\underline{T}}^{(0)\dagger} \cdot \underline{\underline{D}}^{(0)OAO} \underline{\underline{T}}^{(1)} + \underline{\underline{T}}^{(0)\dagger} \cdot \underline{\underline{D}}^{(1)OAO} \underline{\underline{T}}^{(0)} + \underline{\underline{T}}^{(1)\dagger} \underline{\underline{D}}^{(0)OAO} \underline{\underline{T}}^{(0)}, \quad (5.14)$$

must be eliminated. This is precisely the same condition that was utilized above in the diagonalization of $\underline{\underline{D}}^{(OAO)A}$ and $\underline{\underline{D}}^{(OAO)B}$. Hence, the procedure and equations [i.e. Eqs. (5.6), (5.7), (5.8), (5.9), (5.10), (5.11) and (5.12)] can be applied provided the following replacements are made:

$$\underline{\underline{T}} \rightarrow \underline{\underline{U}}, \quad \eta^{(1)} \rightarrow \theta^{(1)}, \quad \underline{\underline{D}}^{(1)OAO} \rightarrow \underline{\underline{D}}^{(1)RO},$$

$$(\lambda_i^{(0)}, \lambda_a^{(0)}) \rightarrow (2, 0). \quad (5.15)$$

Finally, the first-order coefficient matrix is given by the same expression as Eq. (4.14) except for replacing the zero field $\underline{\underline{T}}\,\underline{\underline{U}}$ (now $\underline{\underline{T}}^{(0)} \cdot \underline{\underline{U}}^{(0)}$) by the corresponding first-order quantity (X is independent of the field):

$$\underline{\underline{C}}_{AO}^{(1)RLMO} = \underline{\underline{X}}^{-1} \cdot [\underline{\underline{T}}^{(1)} \underline{\underline{U}}^{(0)} + \underline{\underline{T}}^{(0)} \underline{\underline{U}}^{(1)}]. \quad (5.16)$$

The extension of the field-dependent localization procedure to higher orders is straightforward. In fact, the relationship with the non-canonical CPHF formulation of Karna and Dupuis [2] should be clear from the first-order treatment just given. The equations that we employ are identical to theirs once the appropriate correspondence of variables is made as shown above. Since the same is true for the higher order expressions (the correspondences are unaltered), we do not repeat these results here.

The generalization to dynamic fields also follows the formulation of Karna and Dupuis [2]. They replace the electric field $\mathbf{E}$ (now $\mathbf{E}(0)$) in Eqs. (5.5), (5.6) and (5.7) by

$$\mathbf{E} = \mathbf{E}(0) + \mathbf{E} \cdot (e^{i\omega t} + e^{-i\omega t}) \quad (5.17)$$

for the dynamic case. Taking into account the frequency-dependence there are now three field terms that need to be treated. Thus, the previous expansions in E are replaced by expansions in E, $Ee^{i\omega t}$ and $Ee^{-i\omega t}$, and terms are subsequently grouped according to the power of E as well as the frequency factor. This ultimately leads to the coefficient matrices:

$$\underline{\underline{C}}_{AO}^{(0)RLMO} = \underline{\underline{X}}^{-1} \underline{\underline{T}}^{(0)} \underline{\underline{U}}^{(0)} \quad (5.18)$$

and

$$\underline{\underline{C}}_{AO}^{(1)RLMO}(\pm\omega) = \underline{\underline{X}}^{-1} \cdot \left(\underline{\underline{T}}^{(0)}(\pm\omega)\, \underline{\underline{U}}^{(1)}(0) + \underline{\underline{T}}^{(1)}(0) \cdot \underline{\underline{U}}^{(0)}(\pm\omega) \right) \qquad (5.19)$$

as the zeroth- and first-order expressions. Note that for $\omega = 0$ the above relations reduce to our previous results in Eqs. (4.14) and (5.16), respectively. Proceeding to second order one obtains

$$\underline{\underline{C}}_{AO}^{(2)RLMO}(\omega_1, \omega_2) = \underline{\underline{X}}^{-1} \cdot \Big(\underline{\underline{T}}^{(0)}\, \underline{\underline{U}}^{(2)}(\omega_1, \omega_2)$$

$$+ \underline{\underline{T}}^{(2)}(\omega_1, \omega_2)\, \underline{\underline{U}}^{(0)} + \underline{\underline{T}}^{(1)}(\omega_1)\, \underline{\underline{U}}^{(1)}(\omega_2) + \underline{\underline{T}}^{(1)}(\omega_2)\, \underline{\underline{U}}^{(1)}(\omega_1) \Big)$$
$$(5.20)$$

where ω_1, ω_2 can either be $+\omega$, $-\omega$, or 0.

We are now left to solve the ELG-CPHF equations as presented in KD using the basis defined in Eq. (4.16). In this basis $\underline{S} = \underline{1}$ and the density matrices for different orders are constructed as in KD. After the solution is obtained, the sequence of localization followed by elongation is repeated until the property of interest can be considered sufficiently well converged with respect to system size.

5.4 Examples of Applications of ELG-FF and ELG-CPHF

The elongation method has been used to calculate NLO properties of many interesting materials with a focus on linear and quasilinear polymers. Most of the calculations, thus far, have determined properties in the static limit by means of the finite field (FF) approach. Both semi-empirical and ab initio RHF methods have been employed. Donor/acceptor substituted polydiacetylenes [3], polyimides [4], [Li$^+$[calix[4]pyrrole]Li]$_n$ oligomers (with n up to 15) [5], and meso-meso linked metalloporphyrins (up to 22 units) [6] have all been treated at the RHF/6-31G level. In the latter case an effective core potential (ECP/VDZ) was utilized for the metal atoms Mg, Zn and Ni. A larger 6-311G basis set was employed [7] for RHF calculations on a series of benzo-2,1,3-chalcogen diazole ribbon oligomers (up to 15 units), this time with a core potential (ECP/VDZ) for the chalcogenide atoms (S, Se and Te).

In the following some of the applications will be presented in further detail.

5.4.1 Linear Array of Meso-Meso Linked Porphyrins

With an eye towards potential applications in electronic and optical devices, there has been a considerable long-term effort over the past two decades to synthesize photochemically active polymeric materials that have a π-electron conjugated backbone to which functional groups are attached. Porphyrin arrays constitute one of the most interesting examples because of the wide variety of different structures that can be constructed even from the same porphyrin building unit. These different structures result in a considerable variation of the electronic states as well as the electro-optical properties. Much attention has been paid, in particular, to those cases where the third-order hyperpolarizabilities (γ) have large values [8–11].

On the theoretical side, hyperpolarizability calculations have been performed for porphyrin-based linear chains using the Pariser-Parr-Pople (PPP) valence bond-CI method [12] or semi-empirical molecular orbital theory [13]. There are also numerous experimental studies on highly soluble meso-meso or fused coupled porphyrin arrays (containing, for example, Zn^{II}, Cu^{II}, Ag^{II}, Ni^{II}, or Pd^{II}) that have been synthesized. One example is the work by Osuka et al. [14–19] who measured the electronic, photonic, morphological, and magnetic properties of the synthesized materials. Kim and Osuka [10] have also prepared a variety of porphyrin arrays and performed resonance Raman, as well as other spectroscopic measurements, for chains of various lengths. In their work meso-meso linked porphyrin arrays (denoted Zn by Kim and Osuka) and meso-meso, β–β, β–β triply linked porphyrin arrays (denoted Tn by Kim and Osuka) were reported. They found that the Tn arrays, in particular, show large nonlinear optical responses. From the results of Ref. [10] it appears that longer chains might provide even further enhancements. In that regard, one can readily see how theoretical calculations on longer porphyrin arrays would provide useful information to experimentalists before undertaking tedious synthetic efforts.

The ELG-FF method was applied to porphyrin based π-electron conjugated systems to study their NLO properties. For calibration purposes initial calculations were carried out on the metal-porphyrin polymer Mg-P. Table 5.1 lists HF/6-31G energies and energy differences between elongation and conventional calculations in the presence of an electrostatic field of different strengths. From the table one can see that the energy difference in every case is around 10^{-8} a.u. per atom or even less. Of course, one also has to establish that the parameters obtained from the field-dependent calculations are well-determined and that the parameters obtained from conventional and ELG calculations are very similar. As indicated below we believe that this indeed is the case.

Next we turn to the Zn-P systems that were studied in Ref. [10]. The authors of this work found that the THG γ value for the oligomer at 1,064 nm is almost constant for the Zn structure, regardless of the number of porphyrin units while it increases with length for the Tn structure. In order to explain their results we begin with calculations on the free-base porphyrin, which should behave like the Zn structure assuming that the presence of the metal atom does not make a significant difference (which is the case according to Kim and Osuka [10]). Therefore, we shall report results here for the free base porphyrin. From an inspection of Fig. 5.1 it can be seen that HF/6-31G elongation calculations for the static γ are consistent with the results of Kim and Osuka in that the value per unit saturates quickly and becomes essentially constant at $n = 3 - 4$.

One can see from Fig. 5.1 that CNV SCF calculations can converge till N = 8 at $\theta = 90°$ but only till N = 4 at $\theta = 30°$. Worse SCF convergence for smaller dihedral

Table 5.1 Elongation error in total HF/6-31G finite field energy (in a.u.) compared to conventional calculations for the metal-porphyrin polymer of Mg-P (P = Porphyrin ligand.)

N	E_{tot}(conventional)	E_{tot}(elongation)	ΔE_{tot}	ΔE_{tot} per atom
$E = 0.000$				
4	−3927.0790362013	−3927.0790362023	−1.000E−09	−7.042E−12
5	−4908.5580546429	−4908.5580531303	1.513E−06	8.546E−09
6	−5890.0370723972	−5890.0370692955	3.102E−06	1.463E−08
7	−6871.5160902751	−6871.5160854713	4.804E−06	1.945E−08
8	−7852.9951076808	−7852.9951011258	6.555E−06	2.324E−08
$E = 0.001$				
4	−3927.0803760150	−3927.0803760151	−9.959E−11	−7.013E−13
5	−4908.5598118176	−4908.5598101607	1.657E−06	9.361E−09
6	−5890.0392524007	−5890.0392487430	3.658E−06	1.725E−08
7	−6871.5186959843	−6871.5186903226	5.662E−06	2.292E−08
8	−7852.9981408394	−7852.9981331212	7.718E−06	2.737E−08
$E = -0.001$				
4	−3927.0803760175	−3927.0803760171	3.997E−10	2.815E−12
5	−4908.5598117848	−4908.5598102726	1.512E−06	8.544E−09
6	−5890.0392523938	−5890.0392491635	3.230E−06	1.524E−08
7	−6871.5186959662	−6871.5186910658	4.900E−06	1.984E−08
8	−7852.9981408407	−7852.9981342800	6.561E−06	2.326E−08
$E = 0.002$				
4	−3927.0843981513	−3927.0843981493	2.000E−09	1.408E−11
5	−4908.5650874234	−4908.5650854111	2.012E−06	1.137E−08
6	−5890.0457978979	−5890.0457931756	4.722E−06	2.228E−08
7	−6871.5265207629	−6871.5265130004	7.762E−06	3.143E−08
8	−7853.0072504581	−7853.0072393820	1.108E−05	3.928E−08
$E = -0.002$				
4	−3927.0843981503	−3927.0843981461	4.200E−09	2.958E−11
5	−4908.5650873642	−4908.5650856764	1.688E−06	9.536E−09
6	−5890.0457978985	−5890.0457940926	3.806E−06	1.795E−08
7	−6871.5265206790	−6871.5265145444	6.135E−06	2.484E−08
8	−7853.0072504765	−7853.0072417952	8.681E−06	3.078E−08

The fields are in a.u. and the starting cluster has $N = 4$ units

angles by the CNV method comes from the stronger π-electron delocalization. In contrast, the G-ELG SCF can converge very well regardless of the chain lengths because the number of orbitals in diagonalization is kept as that of the small starting cluster even if the system elongated to large (see Ref. [20]). The SCF cycles on each electric field are shown in Fig. 5.4 for $\theta = 30°$.

From the orbitals and orbital energies determined in an elongation calculation it is possible to extract band structures as they would be in the limit of an infinite periodic system [21, 22] as well as the local density of states (LDOS) [23]. In order to

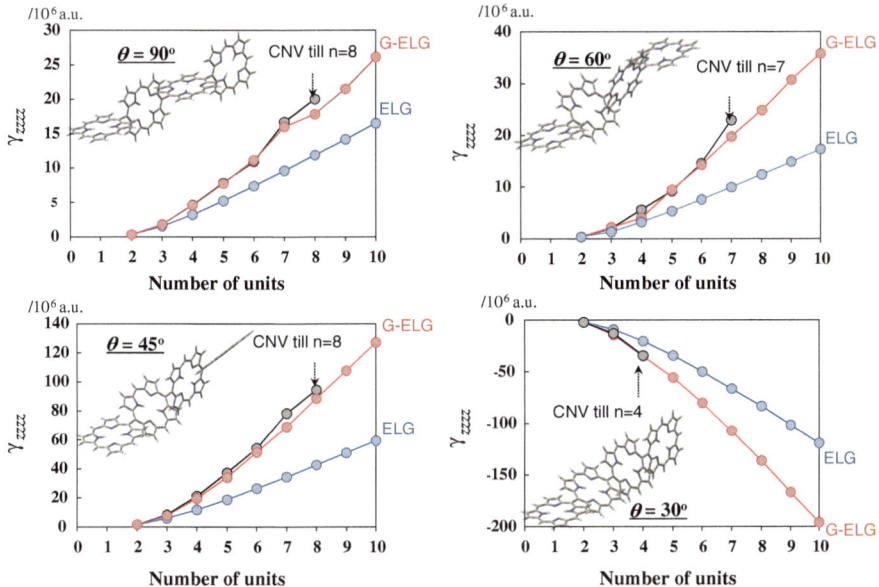

Fig. 5.1 The second order static hyperpolarizability γ_{zzzz} versus number of units in free base porphyrin chains as obtained by the ELG, G-ELG and CNV methods at the HF/6-31G(d,p) level

analyze the differences between the Zn and Tn structures, band structures and LDOS were obtained for three model systems: (a) a Zn porphyrin chain with metal removed (dihedral angle between adjacent units of $90°$), (b) a Zn-containing porphyrin chain (dihedral angle between adjacent units of $90°$), and (c) the zinc-containing fused porphyrin chain Tn. The results shown in Fig. 5.2 were extracted from RHF/6-31G

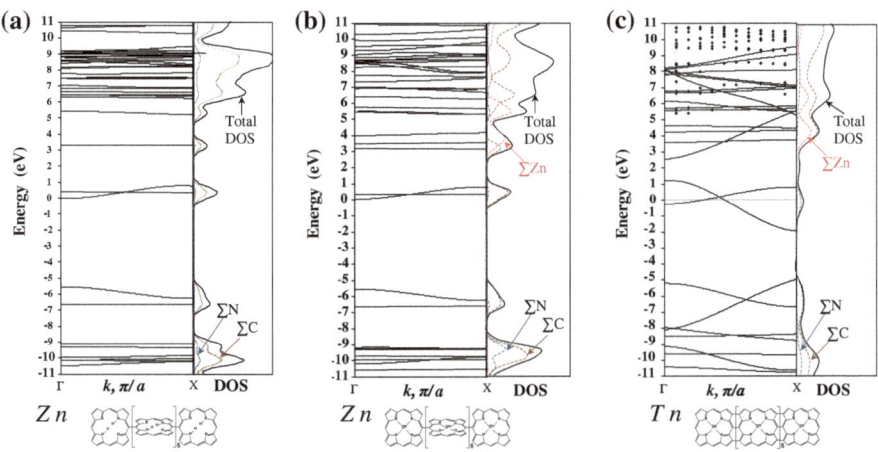

Fig. 5.2 Energy band structures and LDOS from elongation method for the three model porphyrin arrays; **(a)** free-base, **(b)** normal zinc-containing chain and **(c)** zinc-containing fused chain

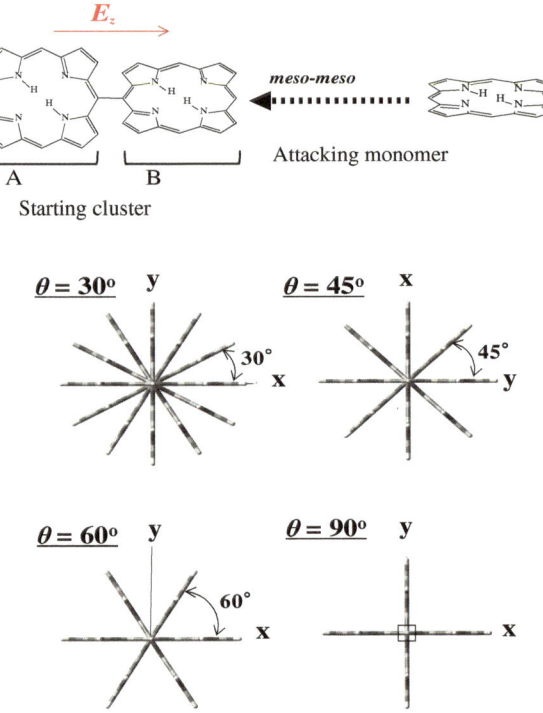

Fig. 5.3 Elongation calculation of model Zn-porphyrin arrays as a function of dihedral angle between neighboring units, depicted in *top* view

calculations using the energy levels and orbitals of the $n = 10$ oligomer. These band structures tell us that the fused system (c) has a smaller band gap than that of the normal Zn-containing porphyrin chain (b), which is consistent with the Tn structure having a larger γ that converges more slowly with chain length. By comparing the LDOS curves of panel (a) without Zn and (b) with Zn, it is also readily confirmed that the Zn bands do not contribute to the states in the vicinity of the Fermi level in (b). Consequently, one expects the NLO properties to be similar for (a) and (b). For the sake of completeness we add that our band structures resemble those obtained using density-functional methods [24, 25] although differences between the two methods lead to differences in some of the details.

As the value of the dihedral angle that was used for the results of Fig. 5.2 is reduced from 90° it may be suggested that the band gap in (b) will be reduced towards that of the Tn structure. In order to study this proposal we performed additional calculations using the G-ELG method (see Sect. 4.6) for the free-base porphyrin at dihedral angles of 90°, 60°, 45°, and 30°; see Fig. 5.3. For even smaller angles the steric hydrogen-hydrogen repulsions between neighboring porphyrin units, which are not present in the Tn structure, make the resulting structures unstable. The Tn type can also be treated using the G-ELG method but we show here only the results for the normal ring systems. The normal system differs from the fused ring system by having four hydrogen atoms between units.

The elongation calculations for the systems of Fig. 5.3 employed a starting cluster of two porphyrin units and, subsequently, a chain elongated by one unit at a time in the presence of an external electric field, E_z, along the chain direction. It is of interest to examine the orbital shift effect as implemented within the G-ELG method. At first we compare the field-free energies obtained by this method with ordinary ELG as well as with conventional calculations (marked CNV) for the free-base porphyrin. Table 5.2 lists the resulting total energies along with energy differences between ELG and CNV and between G-ELG and CNV at the RHF/6-31G(d,p) level. Evidently, errors are reduced about 3 orders of magnitude by the orbital shift technique of the G-ELG method. We note that conventional calculations for a longer chain than $n = 5$ could not be carried out for $\theta = 30°$ because of convergence problems (see more later).

These results suggest that the frozen-orbital reactivated effect could be important for NLO properties of porphyrin chains. Thus, ELG and G-ELG results were also compared for the finite fields of $E_z = \pm 0.001$ and ± 0.002 (the z axis is parallel to the porphyrin chain axis). In this case, we performed calculations at a dihedral angle of 30° and with up to 10 porphyrin units in order to access the behavior for longer chains. For this value the electronic orbitals are very delocalized according to the CNV calculations so that the frozen-orbital reactivated method becomes particularly important. An interesting contrast between ELG and conventional calculations may be seen in Fig. 5.4 where we plot the number of SCF cycles required to reach convergence for the different field strengths and where, for comparison, field-free results are also included. It is seen that the number of SCF cycles needed in the ELG and G-ELG methods is almost constant (less in the case of G-ELG) for all values of n, whereas the CNV method requires an increasing number of cycles as n increases. For the CNV method the SCF procedure did not converge at $n = 6$ for $E_z = 0$, $+ 0.001$, and at $n = 5$ for $E_z = -0.001$, ± 0.002. One reason for such behavior is that the HOMO-LUMO gap in the CNV method approaches zero with an increasing number of units, regardless of whether a field is present or not. This is different for the ELG calculations for which there is a 'local gap' that remains almost constant for the active space used in the ELG-SCF procedure since the active space remains essentially the same size as the overall chain length grows [26]. On the other hand, upon transforming the localized orbitals in the frozen region into delocalized orbitals that are eigenfunctions of the Fock operator for the entire system the gap shows the expected size dependence.

Finally, we discuss the calculated longitudinal component of the second-order static hyperpolarizability γ_{zzzz} for four dihedral angles of the free-base porphyrin as a function of the number of units. The results are presented in Fig. 5.1. They show that the G-ELG method reproduces the conventional curve very well, which is not the case for the original ELG method. For systems with small dihedral angles and thereby very delocalized electrons, γ_{zzzz} cannot be evaluated by the CNV method due to the above-mentioned SCF convergence problem. A more careful analysis of this problem demonstrates that it is due to the high density of states in the vicinity of the HOMO and LUMO, an issue that is relevant even in the absence of a field.

Table 5.2 Total field-free energies (E_{tot} in a.u. per unit) for free-base porphyrin chains using the G-ELG, ELG and conventional (CNV) RHF/6-31G(d,p) methods for four different dihedral angles

Dihedral angle (θ)	$\theta = 90°$ (10)	$\theta = 60°$ (10)	$\theta = 45°$ (10)	$\theta = 30°$ (5)
E_{tot}(ELG)	−982.21700153331	−982.20281905172	−982.16145727217	−982.03292104896
E_{tot}(G-ELG)	−982.21751592066	−982.20329297702	−982.16201033776	−982.03423455460
E_{tot}(CNV)	−982.21751660678	−982.20329331182	−982.16201065376	−982.03423660180
ΔE_{tot}(ELG–CNV)	1.42×10^{-05}	1.31×10^{-05}	1.53×10^{-05}	3.61×10^{-5}
ΔE_{tot}(G-ELG–CNV)	1.90×10^{-08}	9.25×10^{-09}	$9.69\text{x} \times 10^{-09}$	5.62×10^{-08}

The number in parenthesis is the number of units in the porphyrin chain

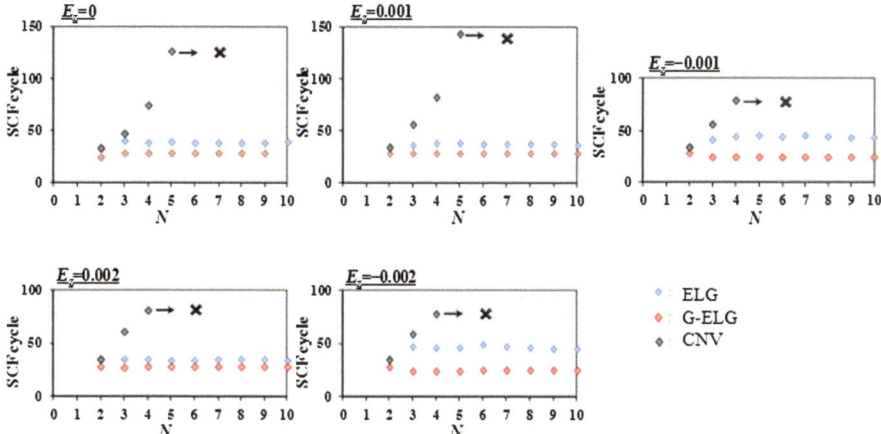

Fig. 5.4 Number of cycles required for SCF convergence using the ELG, G-ELG, and CNV methods for free-base porphyrin chains at a dihedral angle of $\theta = 30°$

It is remarkable that the γ_{zzzz} values for $\theta = 30°$ become increasingly more negative as the number of porphyrin units increases. Nakano et al. [27–32] have suggested that large negative second hyperpolarizabilities can be obtained when the system presents symmetric resonance structures with invertible polarization (SRIP). Thus, it would be worthwhile to further investigate the relationship between delocalization and SRIP, which could be done using the G-ELG method.

The response per unit, i.e., γ_{zzzz}/n, is also of interest. Since, as mentioned above, γ_{zzzz} is negative for some value(s) of the dihedral angle we have chosen to depict $|\gamma_{zzzz}|$ and $|\gamma_{zzzz}|/n$ as functions of n. The results are shown in Fig. 5.5 and demonstrate that even for the largest values of n considered, $|\gamma_{zzzz}|/n$ does not approach a constant value. This implies that the thermodynamic limit is first reached for even larger systems.

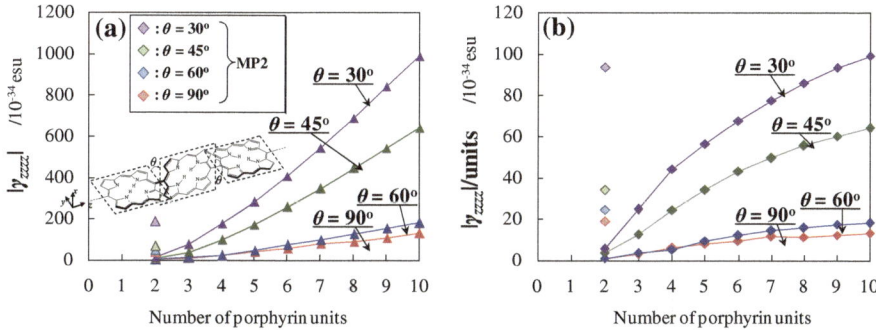

Fig. 5.5 **a** $|\gamma_{zzzz}|$ and **b** $|\gamma_{zzzz}|/n$ versus the number of free base porphyrin units as obtained with the G-ELG method at the HF/6-31G(d,p) level. MP2/6-31G(d,p) $|\gamma_{zzzz}|$ and $|\gamma_{zzzz}|/n$ values obtained by the CNV method are also presented but only for the dimer

The ELG-LMP2 method described in Sect. 4.8 is at the moment of writing being combined with the G-ELG method. However, in a preliminary study MP2/6-31G(d,p) calculations have been carried out for the dimer by means of the CNV method with the purpose of obtaining a preliminary assessment of the effects of electron correlation. In fact, for $\theta = 90°$, correlation effects increase the calculated γ_{zzzz} from 7.5×10^{-35} esu per unit to 1.9×10^{-33} esu per unit. This correlated result is much closer to the value of 10^{-33} esu per unit obtained by Z-scan measurements at 1,064 nm [10]. The fact that another study [33], just five years older, gives values that differ by orders of magnitude indicates the large experimental uncertainty. For a better comparison, of course, a frequency-dependent CPHF/CPKS treatment is needed. In addition, the calculated static γ_{zzzz} values for 10 units are around 100 times larger than those for dimers (~20 times larger per unit). More importantly, the values are still increasing with the number of units at $n = 10$, which may or may not be consistent with the Z-scan experimental measurements (Fig. 5.5).

5.4.2 Water Chain Model System

In this subsection, the water chain model system is exploited to test the ELG-CPHF method. In this model system, the water molecules are arranged in a chain such that O–H bond is along the chain direction and the O–O distance between two neighboring water molecules is set to 2.72 Å. Here the results are obtained by solving the ELG-CPHF equations, whereas all the above results were obtained through the ELG-FF method. For these initial calculations only the static polarizability is considered. In Table 5.3 we report the static polarizability tensor of water chains that are elongated by one unit from an initial structure containing either $N_{st} = 5$ units or $N_{st} = 10$ units. Dividing by the total number of units, N_{fin} (=6 or 11 in the two cases) we

Table 5.3 Comparison between conventional CPHF (CNV) and elongation CPHF (ELG) polarizability tensors per water molecule of a water chain elongated from $N_{st} = 5$ to $N_{fin} = 6$ units, and from $N_{st} = 10$ to $N_{fin} = 11$ units at the HF/STO-3G and HF/6-31G levels

$N_{st} = 5$		α_{xx}	α_{yy}	α_{zy}	α_{zz}
HF/STO-3G	CNV	0.0422	3.2587	−1.4761	5.4865
	ELG	0.0422	3.2587	−1.4759	5.4854
HF/6-31G	CNV	1.4098	5.3936	−0.9500	6.9352
	ELG	1.4099	5.3936	−0.9497	6.9347
$N_{st} = 10$					
HF/STO-3G	CNV	0.0422	3.2536	−1.4767	5.6708
	ELG	0.0422	3.2536	−1.4767	5.6707
HF/6-31G	CNV	1.4098	5.3973	−0.9363	7.0980
	ELG	1.4099	5.3973	−0.9363	7.0979

N_{st} gives the size of the initial structure. The water molecules are lying in the (x, z) plane with the z axis being the chain direction

Table 5.4 Comparison between conventional CPHF (CNV) and elongation CPHF (ELG) polarizability tensors per water molecule of a water chain elongated from $N_{st} = 5$ to $N_{fin} = 10$ water molecules at the HF/STO-3G, HF/6-21G and HF/6-31G levels

		α_{xx}	α_{yy}	α_{zy}	α_{zz}	$\langle \alpha \rangle$
HF/STO-3G	CNV	0.0505	3.9643	−1.8086	4.9078	2.9742
	ELG	0.0505	3.9644	−1.8086	4.9074	2.9742
HF/6-21G	CNV	0.9494	5.7862	−0.8040	6.6246	4.4534
	ELG	0.9494	5.7863	−0.8040	6.6246	4.4534
HF/6-31G	CNV	1.6678	6.3414	−1.3229	7.0317	5.0136
	ELG	1.6678	6.3414	−1.3229	7.0317	5.0136

The water molecules are lying in the (x, z) plane with the z axis being the chain direction

obtain estimates for the polarizability tensor component per water molecule. In both cases the calculations were performed at the RHF/STO-3G and RHF/6-31G levels and the results of the ELG-CPHF method are compared to those of conventional CPHF (marked CNV) calculations. It can be seen that even starting from a cluster of only five water molecules, one obtains very good agreement between conventional and elongation results (maximum difference of 0.0011 a.u. for any component). If one starts from a cluster of ten water molecules, the maximum difference is reduced by an order of magnitude, and one obtains a polarizability tensor that agrees almost exactly with that of a conventional calculation.

Table 5.3 shows that, for a water chain, a starting cluster of five molecules is sufficiently large to obtain reliable ELG-CPHF results for the change in the polarizability tensor due to adding a single molecule. In order to test whether errors accumulate upon making further additions, calculations were carried out for water chains elongated from $N_{st} = 5$ to $N_{fin} = 10$ units (see Table 5.4). It can be seen from the table that even for the largest basis set tested for this model, i.e., HF/6-31G, the largest deviation from the conventional results is less than 0.0005 a.u. This supports the reliability of the ELG-CPHF method for the calculation of NLO properties. Of course, further testing is necessary for more delocalized systems, as well as for higher-orderproperties and frequency-dependence.

References

1. Kirtman, B.: Variational form of Van Vleck degenerate perturbation theory with particular application to electronic structure problems. J. Chem. Phys. **49**, 3890–3894 (1968)
2. Karna, S.P., Dupuis, M.: Frequency dependent nonlinear optical properties of molecules: formulation and implementation in the HONDO program. J. Comput. Chem. **12**, 487–504 (1991)
3. Ohnishi, S., Orimoto, Y., Gu, F.L., Aoki, Y.: Nonlinear optical properties of polydiacetylene with donor-acceptor substitution block. J. Chem. Phys. **127**, 084702 (2007)
4. Yu, G.-T., Chen, W., Gu, F.L., Orimoto, Y., Aoki, Y.: Theoretical study on static (hyper)polarizabilities for polyimide by the elongation finite-field method. Mol. Phys. **107**, 81–87 (2009)

5. Yu, G.-T., Chen, W., Gu, F.L., Aoki, Y.: Theoretical study on nonlinear optical properties of the Li$^+$[calix[4]pyrrole]Li$^-$ dimer, trimer and its polymer with diffuse excess electrons. J. Comput. Chem. **31**, 863–870 (2010)

6. Yan, L.K., Pomogaeva, A., Gu, F.L., Aoki, Y.: Theoretical study on nonlinear optical properties of metalloporphyrin using elongation method. Theor. Chem. Acc. **125**, 511–520 (2010)

7. Pomogaeva, A., Gu, F.L., Imamura, A., Aoki, Y.: Electronic structures and nonlinear optical properties of supramolecular associations of benzo-2,1,3-chalcogendiazoles by the elongation method. Theor. Chem. Acc. **125**, 453–460 (2010)

8. Terazima, M., Hitoshi, S., Osuka, A.: The third-order nonlinear optical properties of porphyrin oligomers. J. Appl. Phys. **81**, 2946–2951 (1997)

9. Kuebler, S.M., Denning, R.G., Anderson, H.L.: Large third-order electronic polarizability of a conjugated porphyrin polymer. J. Am. Chem. Soc. **122**, 339–347 (2000)

10. Kim, D., Osuka, A.: Photophysical properties of directly linked linear porphyrin arrays. J. Phys. Chem. A **107**, 8791–8816 (2003)

11. Ahn, T.K., Kim, K.S., Kim, D.Y., Noh, S.B., Aratani, N., Ikeda, C., Osuka, A., Kim, D.: Relationship between two-photon absorption and the π-conjugation pathway in porphyrin arrays through dihedral angle control. J. Am. Chem. Soc. **128**, 1700–1704 (2006)

12. Matsuzaki, Y., Nogami, A., Tsuda, A., Osuka, A., Tanaka, K.: A theoretical study on the third-order nonlinear optical properties of π-conjugated linear porphyrin arrays. J. Phys. Chem. A **110**, 4888–4899 (2006)

13. Keinan, S., Therien, M.J., Beratan, D.N., Yang, W.: Molecular design of porphyrin-based nonlinear optical materials. J. Phys. Chem. A **112**, 12203–12207 (2008)

14. Aratani, N., Osuka, A., Kim, Y.H., Jeong, D.H., Kim, D.: Extremely long, discrete meso-meso-coupled porphyrin arrays. Angew. Chem. Int. Ed. **39**, 1458–1462 (2000)

15. Kim, Y.H., Jeong, D.H., Kim, D., Jeoung, S.C., Cho, H.S., Kim, S.K., Aratani, N., Osuka, A.: Photophysical properties of long rodlike mesomeso-linked Zinc(II) porphyrins investigated by time-resolved laser spectroscopic methods. J. Am. Chem. Soc. **123**, 76–86 (2001)

16. Aratani, N., Cho, H.S., Ahn, T.K., Cho, S., Kim, D., Sumi, H., Osuka, A.: Efficient excitation energy transfer in long mesomeso linked Zn(II) porphyrin arrays bearing a 5,15-bisphenylethynylated Zn(II) porphyrin acceptor. J. Am. Chem. Soc. **125**, 9668–9681 (2003)

17. Tsuda, A., Nakano, A., Furuta, H., Yamochi, H., Osuka, A.: Doubly meso-β-linked diporphyrins from oxidation of 5,10,15-triaryl-substituted Ni^{II} and Pd^{II}-porphyrins. Angew. Chem. Int. Ed. **39**, 558–561 (2000)

18. Ikeue, T., Furukawa, K., Hata, H., Aratani, N., Shinokubo, H., Kato, T., Osuka, A.: The importance of a $\beta - \beta$ bond for long-range antiferromagnetic coupling in directly linked copper(II) and silver(II) diporphyrins. Angew. Chem. Int. Ed. **44**, 6899–6901 (2005)

19. Tsuda, A., Furuta, H., Osuka, A.: Completely fused diporphyrins and triporphyrin. Angew. Chem. Int. Ed. **39**, 2549–2552 (2000)

20. Aoki, Y., Gu, F.L.: Elongation method for delocalized nano-wires. Prog. Chem. **24**, 886–909 (2012)

21. Pomogaeva, A., Imamura, A., Kirtman, B., Gu, F.L., Aoki, Y.: Band structure built from oligomer calculations. J. Chem. Phys. **128**(7), 074109 (2008)

22. Pomogaeva, A., Springborg, M., Kirtman, B., Gu, F.L., Aoki, Y.: Band structures built by the elongation method. J. Chem. Phys. **130**, 194106 (2009)

23. Aoki, Y., Imamura, A.: Local density of states of aperiodic polymers using the localized orbitals from an ab initio elongation method. J. Chem. Phys. **97**, 8432–8440 (1992)

24. Yamaguchi, Y.: Time-dependent density functional calculations of fully π-conjugated zinc oligoporphyrins. J. Chem. Phys. **117**, 9688–9694 (2002)

25. Pedersen, T.G., Lynge, T.B., Kristensen, P.K., Johansen, P.M.: Theoretical study of conjugated porphyrin polymers. Thin Solid Films **477**, 182–186 (2005)

26. Aoki, Y., Loboda, O., Liu, K., Makowski, M.A., Gu, F.L.: Highly accurate O(N) method for delocalized systems. Theor. Chem. Acc. **130**, 595–608 (2011)

27. Nakano, M., Yamaguchi, K.: A proposal of new organic third-order nonlinear optical compounds. Centrosymmetric systems with large negative third-order hyperpolarizabilities. Chem. Phys. Lett. **206**, 285–292 (1993)

28. Nakano, M., Kiribayashi, S., Yamada, S., Shigemoto, I., Yamaguchi, K.: Theoretical study of the second hyperpolarizabilities of three charged states of pentalene. A consideration of the structure-property correlation for the sensitive second hyperpolarizability. Chem. Phys. Lett. **262**, 66–73 (1996)

29. Nagao, H., Kiribayashi, S., Nakano, M., Yamada, S., Shigemoto, I., Nagao, H., Yamaguchi, K.: Theoretical studies for second hyperpolarizabilities of alternant and condensed-ring conjugated systems II. Synth. Met. **85**, 1163–1164 (1997)

30. Nakano, M., Yamada, S., Yamaguchi, K.: Theoretical studies on second hyperpolarizabilities for cation radical states of tetrathiafulvalene and tetrathiapentalene. Chem. Phys. Lett. **311**, 221–230 (1999)

31. Champagne, B., Botek, E., Nakano, M., Nitta, T., Yamaguchi, K.: Basis set and electron correlation effects on the polarizability and second hyperpolarizability of model open-shell π-systems. J. Chem. Phys. **122**, 114315 (2005)

32. Nakano, M.: Theoretical approaches to the calculation of electric polarizabilities. In: Maroulis, G. (ed.) Atoms, Molecules and Clusters in Electric Fields, pp. 337–404. Imperial College Press, London (2006)

33. Terazima, M., Shimizu, H., Osuka, A.: The third-order nonlinear optical properties of porphyrin oligomers. J. Appl. Phys. **81**, 2946–2951 (1997)

Chapter 6
Future Prospects

There are many chemical systems of interest that are too large to be studied by means of conventional quantum chemistry computational methods. The determination of NLO properties for such systems is particularly challenging because of their complicated nature and the high accuracy requirements for reliable results. One particular approach to such calculations with favorable prospects is the elongation method. Since the original formulation of this method in the early 1990s, it has been steadily evolving in terms of efficiency, accuracy and generality. The most recent generalized version, G-ELG, appears to be capable of handling the strongly delocalized systems that are of particular interest since they exhibit large NLO properties. At the same time the efficiency of earlier versions, based on regional localization and application of cutoff procedures, is preserved even as the accuracy is improved.

The elongation method has been devised to obtain highly accurate total energies. In that regard initial tests of the finite field method for including electrostatic fields (ELG-FF), using G-ELG at the Hartree-Fock level, have proved successful for quasi-1D systems. The extension to two- and three-dimensional systems, which is already available for the field-free case, can be expected shortly. Correlation effects can be very important for NLO properties. We anticipate their being taken into account through application of KS-DFT methods, including those designed to overcome overpolarization due to near-sighted potentials. Alternatively, from the wavefunction point of view, ELG-LMP2 can be employed for this purpose. In field-free calculations it has already been shown that, even with the smallest correlation domain, close agreement of ELG-LMP2 with conventional MP2 calculations (differences of only about 10^{-7} a.u./atom) is obtained.

The study of large finite systems exposed to electrostatic fields may be both challenging and give surprising results. As demonstrated in Chap. 3, the response of any large regular system to such fields will depend upon the surfaces and the surface contributions will have a finite value even for the largest systems. The ability of elongation methods to deal with these surface contributions remains to be studied.

In order to obtain frequency-dependence time-dependent methods must be utilized. The ELG-CPHF procedure has been formulated and successfully tested in a preliminary way. Further testing beyond static linear polarizabilities is required

© The Author(s) 2015

F.L. Gu et al., *Calculations on nonlinear optical properties for large systems*,
SpringerBriefs in Electrical and Magnetic Properties of Atoms,
Molecules, and Clusters, DOI 10.1007/978-3-319-11068-4_6

before general application. As in the case of ELG-FF, correlation may be added in future work by means of KS-DFT response theory. Beyond that one may also envision the development of time-dependent ELG-LMP2.

It has been noted in Chap. 2 that NLO properties may contain substantial, or even dominant, vibrational contributions. The simplest way to estimate the effect of vibrations involves geometry optimization in the presence of an electrostatic field. The energy gradient technique based on RLMOs has already been implemented in the elongation program and the first tests on model systems are promising.

From the above it should be clear that, although the elongation method for NLO and other properties continues to evolve rapidly, the theory remains very much open to future development. Moreover the potential application to nanomaterials of all sorts, that would otherwise be computationally inaccessible, has just begun to be explored.

Glossary

Ab initio methods   Ab initio methods are here used as synonymous with Hartree-Fock methods.

Bloch functions   For an infinite, periodic system the orbitals can be classified according to their properties under the translational symmetry operations. For a translation by the lattice vector $\mathbf{T}$ the functions obey $\psi(\mathbf{r} + \mathbf{T}) = e^{i\mathbf{k}\cdot\mathbf{T}}\psi(\mathbf{r})$. Functions that obey this are called Bloch functions.

Born-Oppenheimer approximation   Within the Born-Oppenheimer approximation, the solutions of the time-independent Schrödinger equation for the complete system consisting of nuclei and electrons are written as products of electronic and vibrational wavefunctions and the kinetic energy of the nuclei is ignored.

Born von Kármán zone (BvK zone)   In a practical calculation for an infinite, periodic system the continuous wavevector $\mathbf{k}$ within the 1st Brillouin zone is replaced by a finite, discrete set of N_k equidistantly spaced points. The corresponding finite unit consisting of N_k unit cells is the BvK zone.

1st Brillouin zone   For an infinite, periodic system most relevant information is obtained by considering those wavevectors $\mathbf{k}$ that lie within a finite volume around $\mathbf{k} = \mathbf{0}$. This volume is the 1st Brillouin zone.

CI method   An electronic structure method for adding correlation effects on top of a Hartree-Fock calculation.

Clamped nucleus approximation   Within the clamped-nucleus approximation it is assumed that the nuclei do not move in response to the external electric field(s).

Coupled-perturbed Hartree-Fock approach (CPHF)   Here, it is used for a perturbation-theoretical approach for including both electronic and nuclear responses to electrostatic fields.

DC Pockels effect (dc-P)   A 2nd order response. In effect, a static field combines with an electric field of the frequency ω resulting in a new field of frequency ω. See Sect. 1.2.1.

Degenerate four-wave mixing (DFWM)   A 3rd order response. In effect, three electric fields of the same frequency ω are combined to a new field of the same frequency ω. The same as intensity-dependent refractive index (IDRI). See Sect. 1.2.2.

© The Author(s) 2015   87
F.L. Gu et al., *Calculations on nonlinear optical properties for large systems*,
SpringerBriefs in Electrical and Magnetic Properties of Atoms,
Molecules, and Clusters, DOI 10.1007/978-3-319-11068-4

Density-functional-theory (DFT) method   An electronic-structure method based on the density-functional-theory of Hohenberg, Kohn, and Sham.

Electric field induced optical rectification (EFIOR)   A 3rd order response. In effect, two electric fields of the same frequency ω are combined with a static electric field resulting in a new static field (with frequency 0). See Sect. 1.2.2.

Electric field induced second harmonic generation (EFISHG or DC-SHG)   A 3rd order response. In effect, two electric fields of the same frequency ω are combined with a static electric field resulting in a new field with frequency 2ω. See Sect. 1.2.2.

Electronic responses   The responses of the electrons to the electric fields. This is in contrast to the vibronic responses.

Electro-optical Kerr effect (EOKE)   A 3rd order response. In effect, one electric field of the frequency ω is combined with two static electric fields resulting in a new field with frequency ω. See Sect. 1.2.2.

Elongation method   An efficient electronic-structure method for large systems that not necessarily consist of identical building blocks.

Finite-field approach   Within the finite-field approach, the responses of the system of interest to electrostatic fields are calculated by studying the changes in the properties for fixed, external fields of different strengths.

Gauge   The presence of an electric field can be described through a vector potential, a scalar potential, or a combination of both. A specific choice defines the gauge. All physical observables will be independent of this choice.

Hartree-Fock method   An electronic-structure method that per definition does not include correlation effects.

Hyperpolarizabilities   These describe the non-linear responses of a molecular material to electric fields. The corresponding properties for a macroscopic material are the susceptibilities. Various processes are discussed in Sect. 1.2.

Infinite, periodic system   A macroscopic system consisting of a large, finite, repeated sequence of regularly placed, identical units can with advantage be treated as being infinite and periodic. For the large, finite system, deviations from the regularity occur only at the boundaries.

Intensity-dependent refractive index (IDRI)   A 3rd order response. In effect, three electric fields of the same frequency ω are combined to a new field of the same frequency ω. The same as degenerate four-wave mixing (DFWM). See Sect. 1.2.2.

Löwdin transformation   A method for transforming one set of non-orthonormal basis functions into a set of orthonormal ones.

Møller-Plesset method (MP2)   A perturbation-theoretical method for the inclusion of correlation effects.

Nonlinear optics   Nonlinear optics (NLO) is the branch of optics that describes the behavior of light in nonlinear media, that is, media in which the dielectric polarization P responds nonlinearly to electric field(s). A summary of the field is given in Sect. 1.1.

Optical rectification (OR) A 2nd order response. In effect, two electric fields of the frequencies ω_1 and ω_2 are combined to a new field of frequency $|\omega_1 - \omega_2|$. See Sect. 1.2.1.

Piezoelectricity Piezoelectricity describes the coupling between mechanical and electric responses of a macroscopic material. Thus, the application of stress can induce a polarization or, equivalently, the application of an external, static voltage can change the spatial dimensions of the sample. The latter is the so called converse piezoelectric effect.

Polarizability This describes the linear response of a molecular material to an electric field. The corresponding property for a macroscopic material is the susceptibility. Various processes are discussed in Sect. 1.2.

Post-HF method Any electronic-structure method that adds correlation effects on top of a Hartree-Fock calculation.

Scalar potential The presence of an electric field can be described through a vector potential, a scalar potential, or a combination of both. A specific choice defines the gauge. All physical observables will be independent of this choice.

Second-harmonic generation (SHG) The prototype of 2nd order responses. In effect, two electric fields of the same frequency ω are combined to a new field of frequency 2ω. See Sect. 1.2.1.

Sum frequency generation (SFG) A 2nd order response. In effect, two electric fields of the frequencies ω_1 and ω_2 are combined to a new field of frequency $\omega_1 + \omega_2$. See Sect. 1.2.1.

Sum-over-states (SOS) approximation A pertubation-theoretical approach to calculate the electronic responses to external electric fields assuming that the nuclei do not contribute. In contrast to the TDHF or TDDFT approaches, the SOS method does not include orbital relaxations.

Susceptibility This describes the responses of a macroscopic material to electric fields. The corresponding properties for molecular systems are the polarizabilities and hyperpolarizabilities. Various processes are discussed in Sect. 1.2.

Third-harmonic generation (THG) The prototype of 3rd order responses. In effect, three electric fields of the same frequency ω are combined to a new field of frequency 3ω. See Sect. 1.2.2.

Thermodynamic limit For sufficiently large systems, any physical observable will be either proportional to or independent of the size of the system. When this size has been reached one has arrived at the thermodynamic limit.

Time-dependent density-functional-theory approach (TDDFT) A perturbation-theoretic approach for calculating the electronic responses to electric fields that includes orbital relaxation (similar to the TDHF approach and in contrast to the SOS approach).

Time-dependent Hartree-Fock approach (TDHF) A perturbation-theoretic approach for calculating the electronic responses to electric fields that includes orbital relaxation (similar to the TDDFT approach and in contrast to the SOS approach).

Vector potential   The presence of an electric field can be described through a vector potential, a scalar potential, or a combination of both. A specific choice defines the gauge. All physical observables will be independent of this choice.

Vibrational responses   The responses of the nuclei to the electric fields. This is in contrast to the electronic responses.

Index

© The Author(s) 2015

F.L. Gu et al., *Calculations on nonlinear optical properties for large systems*, SpringerBriefs in Electrical and Magnetic Properties of Atoms, Molecules, and Clusters, DOI 10.1007/978-3-319-11068-4